HEAVEN AND HELL

The Pope condemns the poor to eternal poverty

Heaven and Hell

The Pope condemns the poor to eternal poverty

Ian Plimer

Connor Court Publishing

PO Box 224W
Ballarat VIC 3350
sales@connorcourt.com
www.connorcourt.com

ISBN: 9781925138801 (pbk)

Cover design by Ian James, photo from the Vatican Photo Service, used with permission

Printed in Australia

ABOUT THE AUTHOR

PROFESSOR IAN PLIMER is Australia's best-known geologist. He is Emeritus Professor of Earth Sciences at the University of Melbourne, where he was Professor and Head of Earth Sciences (1991-2005) after serving at the University of Newcastle (1985-1991) as Professor and Head of Geology. He was Professor of Mining Geology at The University of Adelaide (2006-2012) and in 1991 was also German Research Foundation research professor of ore deposits at the Ludwig Maximilians Universität, München (Germany). He was on the staff of the University of New England, the University of New South Wales and Macquarie University. He has published more than 120 scientific papers on geology and was one of the trinity of editors for the five-volume *Encyclopedia of Geology*. This is his tenth book written for the general public, the best known of which are *Telling lies for God* (Random House), *Milos-Geologic History* (Koan), *A Short History of Planet Earth* (ABC Books), *Heaven and Earth* (Connor Court), *How to get expelled from school* (Connor Court) and *Not for greens* (Connor Court).

He won the Leopold von Buch Plakette (German Geological Society), the Clarke Medal (Royal Society of NSW) and the Sir Willis Connolly Medal (Australasian Institute of Mining and Metallurgy). He is a Fellow of the Australian Academy of Technological Sciences and Engineering and an Honorary Fellow of the Geological Society of London. In 1995, he was Australian Humanist of the Year and later was awarded the Centenary Medal. He was Managing Editor of *Mineralium Deposita*, president of the SGA, president of IAGOD, president of the Australian Geoscience Council and sat on the Earth Sciences Committee of the Australian Research Council for many years. He won the Eureka Prize for the promotion of science, the

Eureka Prize for *A Short History of Planet Earth* and the Michael Daley Prize (now a Eureka Prize) for science broadcasting. He was an advisor to governments and corporations and a regular broadcaster. This book has guaranteed that the author will never be awarded a papal knighthood.

Professor Plimer spent much of his life in the rough and tumble of the zinc-lead-silver mining town of Broken Hill where an interdisciplinary scientific knowledge intertwined with a healthy dose of scepticism and pragmatism are necessary. His time in the outback has introduced him to those who can immediately see the weaknesses of an argument. He is Patron of Lifeline Broken Hill and the Broken Hill Geocentre. He worked for North Broken Hill Ltd and was a director of CBH Resources Ltd. In his post-university career he is proudly a director of a number of listed (Silver City Minerals Ltd, Niuminco Group Ltd, Sun Resources NL, Lakes Oil NL and Kefi Minerals plc) and unlisted Hancock Prospecting companies (Roy Hill Holdings Pty Ltd, Hope Downs Iron Ore Pty Ltd and Queensland Coal Pty Ltd).

A new Broken Hill mineral, plimerite $ZnFe_4(PO_4)_3(OH)_5$, was named in recognition of his contribution to Broken Hill geology. Ironically, plimerite is green and soft. It fractures unevenly, is brittle and insoluble in alcohol. A ground-hunting rainforest spider *Austrotengella plimeri* from the Tweed Range (NSW) has been named in his honour because of his "provocative contributions to issues of climate change". The author would like to think that *Austrotengella plimeri* is poisonous. His dog Benji has never bitten him.

ACKNOWLEDGEMENTS

This book was written at the invitation of my publisher Dr Anthony Cappello (Connor Court Publishing Pty Ltd) because many conservative Catholics were disturbed by and did not agree with the 2015 papal Encyclical *Laudato Si'*. This is my fourth book with Connor Court. When it was known that I was preparing this book, many Catholics contacted me because they were concerned at the Encyclical's message which some thought was confusing and contrary to previous teachings. Others were aware that the Encyclical was a preamble to influence decisions to be made at the 2015 United Nations Climate Change Convention in Paris (COP21/CMP11, 30 November-11 December 2015). Some were appalled at the weak science, the blatant political nature of the Encyclical and the track record of those who advised the Pope.

Because of the timing of the Encyclical and the Paris convention, peer reviewers were under a lot of pressure to read, criticise and review the manuscript such that the book could be released well before the Paris convention. The downstairs study was a mess over winter while this book was being written and I thank my wife Maja for feeding and watering me, discussions on aspects of the book and being a great support. Writing books is no fun. She too has her own books to write and she took time out from her own work to see that *Heaven and Hell* was completed. I thank Bob Besley, John Killick, John Nethery and Kyle Wightman for their efforts in reviewing the book. The four reviewers viewed *Heaven and Hell* with very different eyes and the book has benefited from their fearless criticism, pedantry and knowledge. Any errors are mine.

The book is dedicated to the late Gavin Thomas, former student, colleague and dear friend. After publication of *Telling lies for God* in

1994, GT told me during one of his generous lubricated lunches that the real enemies of civilised Western society were the rabid environmentalists and that my creationist battles were just a training run for the big battles. How right he was.

CONTENTS

1

ALARM BELLS

Nothing changes

In the period 499 to 449 BC, Greece was collapsing, Persia was becoming aggressive and Rome was in disarray. Two-and-a-half thousand years later, it's still the same.

Climate has always changed, the atmosphere has always had a variable carbon dioxide (CO_2) content and vested interests always cry wolf. Again, nothing changes.

The papal Encyclical of 2015 made the same mistake the Church made in 1615 AD by placing the Earth at the centre of the Universe and not the Sun. It is the Sun that drives climate, not the human emissions of CO_2 into the atmosphere of the Earth. Without the Sun there would be no life on Earth. Nothing changes.

As a geologist, I deal with time, space, cycles, planetary processes and small-scale processes. Everything changes, all the time, because the planet is dynamic.

The papal Encyclical

On 24 May 2015, Pope Francis issued his *Encyclical letter Laudato Si' of the Holy Father Francis on care of our common home.*[1] The 184-page letter comprises six chapters and 246 paragraphs of which seven are devoted to pollution and climate change (Paragraphs 20-26, i.e. less than 10% of the Encyclical).[2]

1: http://w2.vatican.va/content/dam/francesco/pdf/encyclicals/documents/papa-francesco_20150524-enciclica-laudato-si_en.pdf
2: Called hereafter *Laudato Si'* in all footnotes

The first chapter (*What is Happening to Our Common Home*) is on the environment and I comment on this Chapter in this book. Chapter Two (*The Gospel of Creation*) is theology and I make no comment.

The remaining four Chapters (*The Human Roots of the Ecological Crisis; Integral Ecology; Five Lines of Approach and Action; and Ecological Education and Spirituality*) are a blend of pseudo-science and green left environmental activism as I show in this book. The Encyclical tries to present a simple understanding and simple solutions to very complex problems and, in many places, is scientifically incorrect.[3]

Contrary to the media hysteria, very little of the Encyclical is about global warming and much is about environmental popularism, economic ideology with a Marxist bent and language that could have been written by Greenpeace. The two previously competing creeds for popular support in the Western world, Christianity and the atheistic belief system of communism, are both declining and the new religion of green left environmentalism is filling the vacuum.

The Pope's Encyclical supports the communist view of the world and promotes the new green left environmental religion. By not challenging the mantras of both communism and environmentalism, the Pope's Encyclical is disturbing or even sacrilegious to many conservative Catholics. Intellectual ignorance, especially on matters of climate and the environment, are dressed up as honouring God.

The Pope is a theological authority. However, he uses the popular rhetorical fallacy *argumentum ad verecundiam*[4] and uses the discredited IPCC as his authority on climate and the environment.

The Encyclical was manna from Heaven for the green left environmental activists and many places with the name Institute were struggling to outdo each other with gushing praise for the Pope. It

---

3: Misquote from H. L. Mencken in "The Divine Afflatus" (*New York Evening Mail*, 16 November 1917): For every complex problem there is a solution that is neat, simple and wrong.
4: Argument by appealing to authority

galvanised the left. Would the same supporters give the Pope support on his views on abortion, same sex marriage and euthanasia?

The Pope should speak for the poor who need cheap reliable base-load coal-fired electricity and potable water. Does the Pope's Encyclical express this care for the poor? I argue here that it is the exact opposite and the Church is siding with the wealthy prophets and profiteers of doom.

The Encyclical yearns for a pre-Industrial Revolution world wherein it is implied that life was simpler, cleaner and happier. As I show later, it was not the case. This naïve Mary Poppins view of the environment never existed. History shows us that lifespan was far lower, famine was common, poverty was universal, human and property rights did not exist, human life was not valued and life was brutal. Humans then had a firm belief in a better after-life, which made the suffering on Earth bearable.

The Encyclical declares that it is designed for all humanity, not just the 1.2 billion Catholics on Earth. I write the first part of this book from Riyadh, the capital of the Kingdom of Saudi Arabia, where it has gained not one column inch in newspapers and has not been mentioned in the local electronic media. It is blatantly political as it is an appeal to Western public opinion before the UN climate change conference in Paris in December 2015.[5]

We know that the Paris conference is coming because climate "scientists" are making scarier and scarier predictions in the preceding months, and the media just love end of the world stories. The Encyclical simply adds fuel to the fire and its release is no coincidence. As with previous climate conferences, the result is predictable. Before any climate conference, delegates should read Mackay's 1841 book *Extraordinary Popular Delusions And The Madness Of Crowds*.[6] Nothing changes.

5: COP21, Paris, November 30-December 11, 2015
6: MacKay, C. 1841: *Extraordinary Popular Delusions And The Madness Of Crowds.* Three Rivers Press

It may well be written for African and South American bishops. It is well-meaning, gentle, naïve and has little bearing on the real world where there is competition, where species kill each other for food, where there is a constant turnover of species by extinction and evolution and where the solution to poverty is not by going back to Gaia. The normal course of climate change is now taken as evidence of impending doom and, naturally, it's all due to human activity.

The Encyclical is surprising in that it embraces the new religion of environmentalism, an Earth worship belief system by city inhabitants that has no history, music, philosophy, charitable or educational role, structure, deep spirituality or understanding of what it is to be human. It fact, the new religion has all the hallmarks of a totalitarian anti-human belief system with instant gratification.

In primitive societies, extreme weather events are explained as punishment from God (or gods) for the sins of man. The Bible, especially the Old Testament, has many such explanations of natural phenomena. In today's period of post-modernist neo-romanticism, extreme weather events are due to the Western world's heedless industrialisation and we will receive our punishment in the form of climate change.

Many commentators have written that the Pope is equating moral virtue with green left political ideology and advocating economic contraction in the West, international agreements on carbon (dioxide) emissions, redistribution of wealth by re-slicing the pie and imposition of sanctions.[7] Neither India nor China will take any notice whatsoever of the Encyclical because they want to bring their people out of poverty and will not agree to wealthy Western hypocrisy, hot air activist breast beating, papal Encyclicals or political games by the energy giants. What China says and what China does are two different things. As I argue later, the papal Encyclical is the road map to push billions of people into poverty.

7: Editorial, *The Australian*, 14 July 2015

Hence this book. It deals with the shortcomings and consequences of the papal Encyclical. When elected, Pope Francis stated he wanted to lead a poor church for the poor yet the consequences of *Laudato Si'* I argue will create more poverty. The Pope does not argue that free markets, personal freedoms, property rights, democracy and cheap and reliable energy will lift billions of people out of poverty (especially in Africa, the Indian sub-continent, Asia and South America).

This happened in the West, primarily because of the coal-driven Industrial Revolution in Western countries during the Enlightenment. The Pope is ignoring the only tried and proven path out of poverty. The Pope is confusing the *"... unfettered greed ..."*[8] with the vibrant economies and the resultant wealth that are needed to provide food, potable drinking water and cheap employment-generating energy. In the past, these changes have never occurred by the application of billions of dollars of international economic aid or redistribution of other people's money. Why should they now?

The Pope invites disagreement and this is exactly the aim of this book. The Pope states:[9] *"We need to develop a new synthesis capable of overcoming the false arguments of recent centuries."* However, the Pope repeats all the well-known false and disproven arguments regarding climate change and the environment. The Encyclical is a dreadful document bereft of science, logic and fact and displays an ideology that can only exist in an economic- and history-free space. With such good records kept at the Vatican, the Encyclical ignores economic history, ignores the history and philosophy of science and ignores the evolution of Western Christianity.

My comments are not a disrespectful attack on the Pope, Christianity or any religion and I acknowledge that for over 2,000 years Christians have attempted to make the world a better place by caring

8: *Laudato Si'*, Paragraph 237
9: *Laudato Si'*, Paragraph 121

and sacrificing for others. And they have. Only a few other religions can make such claims. Selflessness and a focus on stopping poverty and slavery underpin Christianity.

Pope Francis has failed to notice the basic difference between rich and poor countries. The rich countries have successfully developed their fossil fuel resources to create low-cost abundant transport and energy for heating, cooling, cooking, refrigeration, communication, entertainment and employment that has eliminated the grinding burdens of daily living. Humans were once beasts of burden and the Industrial Revolution gave the burden to coal. Humans have spent thousands of years trying not to freeze in cold weather and yet have used ice for many tasks (e.g. keeping fish fresh between the ocean and the kitchen). For almost two centuries, we have been able to have refrigerators in houses run by oil, kerosene, gas or electricity to keep vegetables crisp, to make those necessary ice blocks for drinks and to avoid daily hunting and gathering. Heat rots food, refrigeration preserves food.

At another level, we need a large amount of energy to extract heat from air to make liquid nitrogen (-192°C), which is used in all sorts of medical and scientific applications. MRI scanners need liquid helium (-269°C) and an even larger amount of energy is used to extract heat and keep the helium just above absolute zero (-273.15°C). All this is done primarily by energy from fossil fuels.

His Holiness agonises about excessive consumption in the Western throwaway society,[10] wants slower growth and does not seem to be aware that there is a close correlation between GDP and life expectancy. For example, the World Bank reported that in 1981,[11] 42% of people in the developing world lived on less than a dollar a day. There are now 28% of people in the developing world that

---

10: *Laudato Si'*, Paragraph 22
11: http://documents.worldbank.org/curated/en/1981/01/438420/world-bank-annual-report-1981

live on less than a dollar a day despite a huge population increase.[12] In 1960, the world fed three billion people. Now it feeds more than seven billion. This is unprecedented in the history of time. There is a very simple way to escape poverty. Become wealthy. This takes time.

The Western world

Wealthy countries didn't become wealthy overnight and centuries of free trade, democracy, creativity, resource utilisation and property rights made wealth creation possible. Governments, collectives or international treaties did not create this wealth. Individuals created it. By denying poor countries access to fossil fuels, Pope Francis condemns them to permanent poverty with the associated disease, short longevity and unemployment.

The Western world just didn't suddenly appear, as if by magic. It was a long progressive and retrogressive slog over more than two thousand years through Greek and Roman times, the Dark Ages, the Middle Ages, the Renaissance, the Reformation, the Enlightenment, the Romantic times and now the modern era.

Our thinking processes, religions, literature, art, music, government, freedoms, trade, law, financial systems, engineering and science are a product of this 2,000-year journey. There were great backward periods, such as the Dark Ages. Other cultures have not had the same journey. For me, what defines the difference between Western cultures and those in Asia, Africa and elsewhere is counterpoint. It took about 200 years for Western cultures to evolve from 11th century church music to what is now the technique of combining multiple and different melodies and instruments simultaneously. No other culture has music that uses counterpoint.[13]

12: http://www.worldbank.org/en/about/annual-report
13: Owen, H. 1992: *Modal and tonal counterpoint*. Schirmer Books

The evolution of science

Logic and science became important in ancient Greek times and Aristotle is generally regarded as the father of science. Observation, measurement and a rational explanation of natural phenomena were the start of science although some conclusions can be questioned.

For example, the wearing of amethyst was meant to stop drunkenness and madness. Amethystos is translated as "not drunk." I've conducted an experiment, something that ancient Greeks did not do as part of their science, and I can assure you that the wearing of amethyst does not stop drunkenness. As a trained scientist, I have had to do this experiment many times as part of data replication. This I have done for you, dear reader, as part of enriching your life. Someone has to do it.

Ancient Greece

There are a number of ancient Greek myths about amethyst. I give only one. Dionysus, the god of intoxication and wine, was pursuing Amethystos who, like a good virtuous chaste maiden, refused his advances. She prayed to Artemis to remain chaste and was changed into a white stone that became purple after Dionysus poured wine over the stone. This is not science, it is myth.

A form of democracy evolved in ancient Greece. What we now understand as democracy was certainly not the *demokratia* of Cleisthenes in 509 BC. Only a small segment of the community could participate and these were second-generation Athenian males over the age of 18.[14] Democracy in the Western world has many similarities (e.g. the division of power). The US system is different from that of Canada, UK, Germany, Australia and many other jurisdictions. Western democracy has optional, preferential and compulsory voting systems, two party, multi-party or a coalition of parties and heads of state plus executive power. Western democracies are a long way from

14: Holland, T. 2013: *Herodotus: The histories*. Penguin Classics

the original Greek *demokratia*. Countries that claim to be a Democratic Peoples' Republic normally are totalitarian, undemocratic or a family feudal system.

Roman law

Long before the Roman Republic was established, Roman law had evolved through the generations as an inherited aspect of society.[15] In 451 BC, a committee was established to write down the law.[16] No new law was created; customary law was just written down. The growth of the Roman Republic over a large area led to the law of nations (*ius gentium*) and natural law (*ius naturale*). *Gentium* was the body of law based on reasoning that society and humans were understood to live by and share whereas *naturale* was based on the principles shared by all living creatures.

The Justinian compilation of legal codes was lost after the fall of Rome, was revived in Bologna in the 11th century, and spread throughout Europe. Although the Western Roman Empire collapsed in 476 AD, later systems of law in Western societies borrowed heavily from the body of civil law (*corpus iuris civilis*) underpinned by the Justinian legal code of 534 AD that evolved in the Eastern Roman Empire.

Lights out in Dark Ages

After the decline of the Roman Empire when the light of Rome didn't shine, there was a cultural, economic and social decline. This was the Dark Ages. Furthermore, it was a period of natural cooling, greatly variable weather and economic contraction when crops failed resulting in starvation, disease, economic collapse and war. It was a terrible time to live. Don't give me the good old days. They weren't. Many of the advances of Greek and Roman times were lost although the scientific method stayed alive in the Arab world.

15: https://www.law.berkeley.edu/library/robbins/RomanLegalTradition.html
16: The Twelve Tables

In 11th century Iraq, Alhazen wrote about the scientific method and argued that the seeker of truth does not place his faith in consensus. Instead, by using hard-won scientific knowledge, the scholar tries to verify this knowledge. Alhazen doubted the work of Ptolemy and *inter alia* stated:[17] *"The road to the truth is long and hard: but that is the road we must follow."* Such thinking did not see the light of day for centuries in the Western world.

One measure of the economic productivity and dissemination of knowledge in the Dark Ages is the copying of manuscripts.[18] In the 6th, 7th and 8th centuries, almost no works were copied, there was a slight rise in the 9th, 10th and 11th centuries and from the 12th century onwards manuscript copying increased exponentially in the Middle Ages, Renaissance, Reformation and Enlightenment periods. Although the Chinese invented movable type printing in the 11th century AD, in the Western world movable type printing was invented in 1440 AD by Johannes Gutenberg and mass publication of books led to widespread reading, better education and enlightenment.

Warm times

In the Early Middle Ages, there was social disruption with depopulation, de-urbanisation, invasion and human diasporas and new kingdoms came and went. In the High Middle Ages, the agricultural and technological output of Europe greatly increased because of a natural period of warming. There were no CO_2-emitting industries that made the planet warmer in what is called the Medieval Warming. It was natural.

Green left environmental activists need to understand that there have been many natural cycles of warmings (e.g. Minoan, Greek-

---

17: http://www.africaresource.com/rasta/sesostris-the-great-the-egyptian-hercules/ wise-words-of-alhazen-the-moorish-master-and-founder-of-the-scientific-method/
18: Bruingh, Eltjo and van Zanden, Jan Luiten 2009: Charting the 'Rise of the West': Manuscripts and printed books in Europe, a long-term perspective from the sixth through 18th centuries. *Jour. Econ. History* 69: 409-445

Roman, Medieval and Modern) and coolings (Dark Ages, Little Ice Ages), that previous warm times were warmer than at present and the world didn't end, sea level did not rise alarmingly, there was no increased storm or extreme weather events and populations actually thrived in the warmer times. However, rather than trying to come to a deep understanding about climate and history, the green left environmental activists just simply remove the Medieval Warming from the record.[19] When people rewrite history, motives can only be sinister.

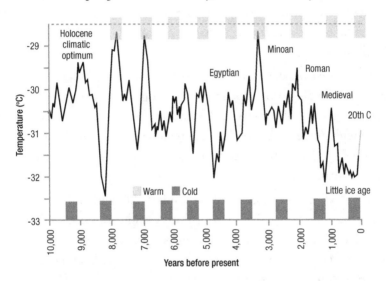

Figure 1: Greenland GISP2 ice core interglacial temperature vs time showing Holocene climate optimum, Egyptian, Minoan, Roman, Medieval and Modern warmings.[20] Note that for most of the last 10,000 years, the Earth has been warmer than now and note that in the current interglacial the temperature trend is downwards.

19: Klein, Naomi 2014: *This changes everything: Capitalism vs. The climate.* Simon and Schuster

20: Alley, R. B. 2000: The younger Dryas cold interval as viewed from central Greenland. *Jour. Quat. Sci. Revs* 19: 213-226

Unless the well recorded past events of natural global warming can be explained rationally, we should be very careful to claim that the warmer times we enjoy today are due to our activities and not due to natural climate cycles. The great humbling of time and scale that geology teaches us gives relevance to changes that occur today. What we see today is nothing compared with past great events. It helps not to be narcissistic or be blessed with human arrogance. It was the cold times that killed off humans, killed plants and animals and resulted in desertification. In the geological past, it was cold times that reduced the number of species. We see the same today in species distribution.

Some 90% of all species live in the tropics with fewer than 1% at the poles.[21] This is no surprise. As a geologist, when I am examining a fossiliferous sedimentary rock outcrop, I do a quick species count as a rough guide as to whether this material was originally deposited in a high or low latitude setting. Later palaeomagnetic studies can give an accurate assessment.

To be alive on planet Earth in the late 20th and early 21st centuries is winning the global climate lottery and, in periods of natural warming, there is longevity, population growth, economic growth, less starvation and more rule of law. It has always been far better to live in warmer times than colder times. It still is.

In the Middle Ages in Greenland, wheat and barley were cultivated, cattle grazed the grasslands and graves could be dug because the permafrost had disappeared. The Vikings were the first to feel the Medieval Warming and started to expand their territories, fisheries and agricultural lands to the UK, Europe and Russia. Europe was warmer than now, crops were planted at altitudes and latitudes where crops could not thrive today, there were fewer storms, tree

---

21: Brown, J. H. 2014: Why are there so many species in the tropics. *Jour Biogeography* 41: 8-22

lines in the Alps moved up slope and glaciers retreated. This gave a period of great economic stability, population increase and economic growth.

For hundreds of years, crops did not fail, there were often two harvests a season in the Northern Hemisphere, wealth was accumulated, excess wealth was spent building universities, cathedrals (e.g. Chartres, France) and monasteries as part of trying to convert pagan Europe to Christianity, great crusades to the Holy Lands were undertaken, peasants lived in feudal villages that owed their lives to nobility whereas nobility owned land in return for military service, kings headed centralised nation-states and crime and violence decreased. For those of us in the Anglosphere, it is relevant that a pretty boring field adjacent to the upper Thames River (Runnymede) was where the *Magna Carta* was signed. King John's grant of *Magna Carta* on 16 June 1215 AD was a revolutionary advancement in the law whereby the king could be bound by the law thereby establishing a clear formal recognition of the rule of law. Taxation became prescribed rather than random.

The stable warm prosperous times of the Medieval Warming led to great theological and philosophical advances (e.g. Thomas Aquinas 1225-1274 AD), art (Giotto 1266-1337 AD) and poetry (Dante 1265-1321 AD, Petrarch 1304-1374 AD, Boccaccio 1313-1375 AD, Chaucer 1343-1400 AD). This was the start of the Renaissance.

The Renaissance

The Renaissance (14th-17th centuries AD) was the bridge between the Middle Ages and modern times. It started as a cultural movement in Italy and spread throughout Europe. All people in the West are beneficiaries of the Renaissance. It took two decades for the climate to change from the Medieval Warming to the Little Ice Age between 1280 and 1303 AD. Major seaways were covered in ice (e.g. Gulf of Bothnia, 1303 AD), the planet cooled, crops failed, populations starved and the weakened ones succumbed to the black death.[22]

The first well-recorded global pandemic in 542 AD (Plague of Justinian) resulted in the death of up to 50 million people and the second was the black death between 1347 and 1350 AD when a third of all Europeans died prematurely. In some cities up to 60% of people died from the disease. The plague has not left us. Since the black death, there have been localised outbreaks of the plague in Europe in the 17th and 18th centuries and in China and India in the 19th century.

In the early 20th century, there were even multiple localised occurrences of the plague in Australia. There are still reports of isolated cases. For example, twice in 2015 a Yosemite National Park (USA) camping ground was closed due to two people contracting the plague, probably from squirrels. There were another two outbreaks in Colorado and one adult died.[23]

The Little Ice Age at the beginning of the Renaissance led to civil unrest, depopulation, economic collapse, peasant revolts and conflicts in the church. The black death and associated massive depopulation led to renewed thinking about one's life on Earth, spirituality and the afterlife. This may have stimulated a boom in Renaissance religious paintings.

It is not known why the Renaissance started in Florence where many major artists thrived (Leonardo da Vinci, Sandro Botticelli and Michelangelo Buonarroti to name a few). Not only did art thrive but also new and exciting academic disciplines such as poetry, grammar, history, moral philosophy and rhetoric thrived as did mathematics, natural philosophy (i.e. science) and astronomy. The process for

---

22: Bubonic plague was carried by fleas (*Yersinia pestis*) on ships that visited Asian ports and spread through Europe because of poor sanitation and a lack of reticulated potable water.

23: http://edition.cnn.com/2015/08/18/health/yosemite-plague/

scientifc discovery, the scientific method, arose and mathematics and empirical evidence developed as methods of understanding nature. Polyphony and counterpoint arose from somewhat colourless church music, mainly chants. Accountancy was invented.

The Renaissance began in times of religious turmoil, three men concurrently claimed to be the Pope and it was not until 1512 AD with the Fifth Lateran Council that the papacy emerged as the supreme authority on ecclesiastical matters.

The Italian Renaissance moved northwards and there was innovation in architecture, art (e.g. Albrecht Dürer, Pieter Bruegel, Hieronymus Bosch, El Greco), literature (e.g. François Rabelais, Miguel de Cervantes), poetry, ballet (e.g. Caterina de' Medici) and music. In England, the Renaissance led to writers (e.g. William Shakespeare, Francis Bacon and John Milton) and composers such as Thomas Tallis and William Byrd. The Renaissance and Reformation with their creativity, innovation, questioning of religion and freer thinking sowed the seeds for the age of Enlightenment (17-18th centuries). Printing of pamphlets and books allowed new ideas to spread like wildfires.

The Reformation

The publication of the 95 Theses by Martin Luther (1483 to 1546 AD) in 1517 AD led to the Reformation. New ideas, pamphlets, books, printed music and heresies were able to spread quickly because of the movable type printing press, the invention of better paper and inks and the availability of numerous cheap printed books. Circulation of printed material allowed authority to be challenged. Fertile grounds of dissent allowed Martin Luther (and John Calvin and Henry VIII) to light a grass fire that became a bushfire. Luther was excommunicated and translated the Bible into German. This took power away from the priests.

The Reformation splintered Catholic Europe, created the foundations for modern Europe and led to the Church of England. The Reformation sought the redistribution of political and religious power and triggered wars, persecution and a counter-Reformation in response to Protestantism. In the counter-Reformation, the Catholic Church became more spiritual, literate and educated. New religious orders appeared (e.g. Jesuits) and there was a revival of older orders. Inquisitions in Spain and Italy fought Protestant apostasy and heresy. The Reformation in Germany finished with the Peace of Augsburg, which allowed the coexistence of Protestantism and Catholicism. There were peasants' revolts and the Treaty of Westphalia in 1648 AD ended the Thirty Years' War, which cost Germany about 40% of its population. In north Germany, Scandinavia and the Baltic States, Lutheranism became the state religion.

It was not all bad. Lutheran church music, universities and baroque art thrived and the Dutch Calvinist merchants spread capitalism. Because of J. S. Bach (1685 to 1750 AD), music became far more technical with harmonic scales. Bach's *"Das Wohltemperierte Klavier"* was a collection of 48 pairs of preludes and fugues in each of the 12 major and minor keys. This led to the shift from Pythagorean scales to the well-tempered scale that avoided some music sounding dreadful in one key and wonderful in another.[24] I would argue today that the modern jungle music has shifted to a disharmonic cacophony, another sign of the current post-modernist neo-romanticism.

The Enlightenment

In the Western world, we are all beneficiaries of the Age of Enlightenment in the 17th and 18th centuries. Coffee houses, private sa-

---

24: Why does this clown Plimer write about counterpoint, Bach, scales and well-tempered music. It is because this book is written to the web-streamed music of Radio Swiss Classic. Check it out on www.radioswissclassic.ch, it will lift your day. Forget the ABC, even the music sounds red.

lons, debating societies and lodges were places of vibrant discussion. This period is sometimes called the Age of Reason because philosophers[25] such as Bacon, Descartes, Locke, Spinoza, Newton, Voltaire, Hume and Kant had a great influence on everyday life. Kings and other rulers sought intellectuals, enacted previously unthinkable social reforms, developed more tolerance and supported the scientific revolution started by Newton.

Knowledge was previously constrained to religious texts and many new great encyclopediae, dictionaries and thoughts were published and widely circulated. Reason challenged established ideas, churches and monarchies (which led to the French Revolution). Universities became places of vibrant thinking and a diversity of thought. The scientific method was invented, great discoveries were made and new industries developed. Many French universities remained places of tradition and changed little during the Enlightenment despite a significant French input into the Enlightenment. The public was very much involved with reading societies, educational institutes, evolution of new political parties and the growth of new nations.

Geographic exploration, combined with natural science, astronomy and anthropology were common (e.g. HMS *Endeavour*, HMS *Beagle*, *La Recherche*) in what was an age of discovery. Scientific societies started and thrived and great scientific discoveries were made.

25: Key great intellectual figures were: Francis Bacon (1562-1626 AD), Thomas Hobbes (1588-1679 AD), René Descartes (1596-1650 AD), John Locke (1632-1704 AD), Baruch Spinoza (1632-1677 AD), Robert Hooke (1635-1703 AD), Isaac Newton (1642-1727), Gottfried Leibniz (1646-1711 AD), Emanuel Swedenborg (1688-1772 AD), François-Marie Arouet Voltaire (1694-1778 AD), Benjamin Franklin (1706-1790 AD), G. L. Buffon (1707-1788 AD), Carl von Linné (1707-1778 AD), David Hume (1711-1776 AD), Jean-Jacques Rousseau (1712-1778 AD), Denis Diderot (1713-1784 AD), Immanuel Kant (1724-1804 AD), Adam Smith (1723-1790 AD), Edmund Burke (1729-1797 AD), Luigi Galvani (1737-1798 AD), Edward Gibbon (1737-1794 AD), Thomas Jefferson (1743-1826 AD), Antoine Lavoisier (1743-1794 AD) and Johann Wolfgang von Goethe (1749-1832 AD).

Many scientific ideas were discarded on the basis of evidence and reason. Freemasonry may have had a significant effect on the Enlightenment because many great names of the Enlightenment were Freemasons (e.g. Diderot, Voltaire, Horace Walpole, Robert Walpole, Mozart, Goethe, Frederick the Great, Benjamin Franklin and George Washington). The book industry boomed and music grew with great European composers experimenting with music (Bach, Mozart, Haydn).

The Industrial Revolution

The first Industrial Revolution was a direct result of the Enlightenment. It was 2,000 years in the making. Coal, oil and gas are very efficient natural means of storing the energy from the Sun. Humans invented a process whereby pre-historic sunlight was unlocked from coal and used to make the world a better place. Suddenly, humans were no longer beasts of burden.

Where did these fossil fuels come from? Coal was originally plant material that, over time, was deeply buried and converted from peat to brown coal and eventually to black coal by temperature and pressure. The energy density increased during this process. The original plant material extracted CO_2 from the atmosphere by photosynthesis. Burning of this coal puts the CO_2 back into the atmosphere where the CO_2 can again be sequestered into life and sediments.

The plants that formed peat grew in a cold climate that had a high atmospheric CO_2 content yet we are told that a higher atmospheric CO_2 content creates global warming. We hear the argument that by burning coal, we are putting CO_2 back into the atmosphere at a rate far faster than nature puts CO_2 into the atmosphere hence we are upsetting natural systems. Wrong. One decent volcanic fart can put more CO_2 into the atmosphere in a few days than all humans have put into the atmosphere over centuries.

What happens to this volcanic CO_2 in the atmosphere? It is sequestered into plants, the oceans and marine life. Oil forms from minute floating organisms in the ocean that extract CO_2 from seawater,[26] die and accumulate in sea floor sediments. With time, temperature and pressure, these organisms are converted into crude oil and gas. If the oil and gas remain in the sediment or move only a short distance (i.e. tight oil, tight gas), oil and gas can be extracted by fracking. Oil and gas that migrate through a pile of sediments can be trapped in geological structures (conventional oil and gas). Burning of oil and gas puts the CO_2 back into the atmosphere to again be used by life or dissolved in seawater. Again, one volcanic fart over a few days can put far more CO_2 into the atmosphere than burning all our oil and gas. Perspective is not a strong card in the green left environmentalist pack.

Although the Romans used coal in England, it was not until the 18th century in the Industrial Revolution that the English found that coal burned cleaner and hotter than wood and charcoal. Coal powered the new technologies for steel making, textiles and transport invented in the Industrial Revolution and the widespread use of coal saved the forests in the UK, Europe and USA in the 18th, 19th and 20th centuries. The use and costs of fossil fuels diverted humans from clear felling forests for wood for cooking, heating, building, glass making and smelting; killing whales and seals for lamp oil; and damming every brook, stream and river for energy and irrigation.

During the Industrial Revolution, millions of people were able to escape from the slavery of working on the land, gripping poverty and move to a city. People had a future and did not necessarily work in the identical job to their fathers and grandfathers. Slavery under a landlord disappeared and the middle class appeared, grew, travelled and became more educated.

Great environmental benefits were gained by changing to fos-

26: CO_2 in seawater is as dissolved CO_2, dissolved CO_2-bearing air, HCO_3^- and CO_3^{2-}

sil fuels during and after the Industrial Revolution. China and India were the most powerful and richest countries from the time of Jesus until the Industrial Revolution. The USA, UK, Europe and Japan then experienced stupendous growth.[27] It is only in the last 50 years we have seen significant growth in Africa, China and India.

Although life was tough, it got better thanks to coal and the life expectancy in England doubled. Coal allowed for stunning increases in productivity. In the mid 18th century in Europe, the forests were rapidly disappearing and coal reversed this trend. Coal greatly reduced pollution caused by cooking and heating with wood, twigs, leaves and dung. Coal allowed the large scale smelting of metals. Coal paved the way for modern agriculture, medicine, commerce and freedom of movement on a scale never imagined before the Industrial Revolution. Coal allowed ordinary people leisure time, made the growth of democratic institutions possible and led to the abolition of slavery. People lived longer, ate better and their purchasing power increased year-in year-out.

Coal gave us cheap electricity in the 20th century and, with technological inventions such as fluidised bed combustion, pollution was decreased. Some 1.3 billion people on Earth still have no access to cheap coal-fired electricity and live in misery. On moral grounds, the Pope cannot deny these folk the benefit of coal. Coal led to the flowering of culture and the growth of the Church of England. This is the same Church of England that now wants to divest itself of investments in coal, has declared a propaganda war on coal from the pulpit and is committing a dreadful crime against the world's poor.

Since 1800, the global GDP[28] has increased by a factor of 10, life

---

27: The economic history of the last 2,000 years in 1 little graph. *The Atlantic*, 19 June 2012 and 22 June 2012

28 Fascinating new graph shows the 'Economic history of the world since Jesus'. *Daily Mail*, 24 June 2012

expectancy has doubled, infant mortality is down by a factor of six and global energy use has increased by 26 times.[29] In the US, the last 215 years have seen a 9,000% increase in the value of goods and services available to the average American, almost all of which are made with, made of, powered by or propelled by fossil fuels.[30] Today, air, water and soil pollution are declining in the major industrialised nations. Nations that use the most energy and fossil fuels have the best air, water and soil quality and the best methods for disposal of waste. Now that incomes have risen, some developing nations are also able to reduce air, water and soil pollution. The Pope should have been advised that increased fossil fuel consumption keeps bringing people out of poverty, not the inverse.[31] The Pope suggests:[32]

> *We know that technology based on the use of highly polluting fossil fuels – especially coal, but also oil, and to a lesser degree, gas – needs to be progressively replaced without delay.*

This reads like a glib unsubstantiated propaganda statement from a green left environmental activist document. Maybe it is? In the papal Encyclical there is no recognition that coal brought millions of people out of poverty and there is not one mention of the Industrial Revolution. Has this great change in the Western world been wiped from history because it does not fit the narrative?

In the "good old days", the *Mayflower* and its 102 pilgrim passengers and crew took 120 days to travel from Plymouth (England) to what was to be Cape Cod (USA) in 1620 AD. Two passengers didn't make it because they died *en route*. Now some 25,000 commercial aeroplanes carry nine million people each day for a total of 15 billion kilometres, thousands of people travel the seven-hour UK-US route by air every day. Many passengers don't make it, only because

29: www.worldenergyoutlook.org
30: http://www.deirdremccloskey.org/docs/pdf/IndiaPaperMcCloskey.pdf
31: *Laudato Si'*, Paragraphs 23 and 26
32: *Laudato Si'*, Paragraph 165

they were late and missed the flight. In the "good old days", there was national and international trade[33] for the same reasons there is trade today.

Modern international trade uses 90,000 bunker fuel-burning vessels, mostly container ships, that burn about half a trillion tonnes per year of fossil fuels and emit CO_2 and sulphur gases. In addition, there is inestimable industrial, automobile and heavy haulage using fossil fuels just to bring food and consumer products to your door. Except for Russia, no country has all the commodities it needs within its political boundaries.

Because of this, some countries such as the USA have strategic stockpiles of commodities they cannot produce or cannot readily obtain if the seaways are closed during a period of conflict. Today many perishable foods, flowers and high unit value commodities are transported by air. In the Encyclical there is absolutely no mention of trade (apart from the drug trade[34] and illegal trade in flora and fauna[35]), travel or commodity stockpiles and one wonders how Rome thinks the world operates.

However, a large tonnage of commodities can be shifted by air if necessary. For example, in the Berlin blockade (1 April 1948 to 12 May 1949), the citizens of West Berlin had no coal for electricity, cooking and heating because of a Soviet road and rail blockade. The two million people of West Berlin needed 1,534 tons of food and medicines and 3,475 tons of coal and liquid fuels a day. The solution was simple but expensive. At the height of the blockade, nearly 9,000 tonnes of coal and food was flown in from West Germany to Berlin day and night mainly by DC-3 and DC-47 aircraft. Also used

---

33: Oudbashi, O. *et al.* 2012: Bronze in archaeology: A review of archaeometallurgy of bronze in ancient Iran. In: *Copper alloys – early applications and current performance-enhancing processes* (Ed Collini, L.) InTech
34: *Laudato Si'*, Paragraphs 123 and 197
35: *Laudato Si'*, Paragraph 168

were DC-4, DC-54 Skymaster, Avro Yorks, Handley Page Hastings, C-82 Packets, C-74 Globemaster, YC-97A Stratofreighter and Lockheed Super Constellation aircraft.[36] Even Sunderlands and Catalinas landed in the harbour to deliver corrosive salt that could not be carried in other aircraft. A total of 692 aircraft were used in the airlift and more than 100 belonged to civilian operators.

Large tonnages were delivered by the US Air Force (1,783,575 tons), RAF (541,937 tons), RAAF (7,968 tonnes and 6,964 passengers), Royal Canadian Air Force, Royal New Zealand Air Force and the South African Air Force. More coal and food were delivered daily than had previously been transported by rail. Why write so much about the Berlin blockade that happened nearly 70 years ago? Because it shows that humans have the ability to solve very difficult problems and that coal and other fossils fuels are essential for modern life. Green left political activists need to learn that all environmental problems can be solved by science and technology and that ideology, catastrophism and doom and gloom have never solved a problem.

Some commodities were once traded extraordinary distances by camel, donkey, horse, wagon, sail and shank's pony (or mare). For example, there are chemical fingerprints of Chinese tin in Persian bronze[37] suggesting thousands of kilometres of overland travel. Trade has expanded. In 1900, world exports had a value of $US10 billion (in today's dollars). In 2013, it was $US18 trillion and rising. Each day fossil fuels transport 100 million tonnes of freight. If fossil fuels were not used, there would be little trade, massive poverty

---

36: Besides music, the author has an interest in the Historical Aircraft Restoration Society (HARS) at Albion Park, NSW. Go there and look at a few of the planes listed that have been fully restored by volunteers. HARS has one of the two surviving Lockheed Super Constellations in the world.

37: Snoek, W. *et al.* 1999: Application of Pb isotope geochemistry to the study of the corrosion products of archaeological artifacts to constrain provenance. *Jour. Geochem. Explor.* 66, 421-425

and shortages of basic commodities. Do we really want to stop using fossil fuels?

In the late 19th century, 500 tonnes of horse manure was excreted on streets and in stables in New York each day.[38] Dead horses were left rotting on the streets, disease was rampant causing 20,000 deaths. The smell was overwhelming. In 1890, the average New Yorker took almost 300 horse car rides a year. The problems were solved by human ingenuity, not by governments or carping environmentalists. The motorcar was invented and welcomed in New York on environmental grounds. It was science, technology and fossil fuels that reduced pollution in New York. In today's world, everything the Pope uses results from mining, fossil fuels or intense agriculture for the massive production of food; we live in a golden age of low-cost energy.

The development of modern science went hand-in-hand with the rise of modern capitalism during the Industrial Revolution. Steelworks, cotton mills, potteries, mines, railways, canals, roads and mechanisation were all constructed by private enterprise. Individuals who had a burning curiosity undertook scientific research. There was no official government policy on scientific research or the direction of science. Free market scientific research led to individuals making profound scientific discoveries in the Enlightenment by following their curiosity and passion. The government did not fund science. Maybe that's why there were so many great scientific discoveries and practical inventions at that time.

Probably the greatest advance came from patents. Patents allowed inventors to profit from their hard work and creativity and intellectual property rights became as important as terrestrial property rights. The role of government was to enforce patents, maintain a healthy legal and commercial environment and to protect trade and

---

38: http://www.nyhistory.org/community/horse-manure

the Empire. Scientists are now paid by government and much of what commonly passes as "science" is taxpayer-funded environmental activism. Science today is driven by the next research grant and curiosity-based research rarely gets funded. Ask me, I've been there.

In the age of Enlightenment, there was an explosion of scientific discovery, men's minds escaped from the shackles of subservience to authority (political and ecclesiastical) and technological development exploited the achievements of science (e.g. steam locomotive, electric light, telephone). The 18th and 19th centuries were a time of great optimism, the 20th century had a crippling depression and two World Wars which exposed the darker side of humanity and in the peaceful 21st century of affluence, doom and gloom have captured those who claim that they are the progressives of the intellectual classes. Even Pope Francis states:[39]

We may be leaving to coming generations debris, desolation and filth.

and

The earth, our home, is beginning to look like an immense pile of filth. In many parts of the planet, the elderly lament that once beautiful landscapes are now covered with rubbish.[40]

His Holiness is wrong. Life is better. Compared to former times, we live longer; have better health, better education, more food and more assistance from science and technology.[41] There is less filth in the developed world. Some of us remember how filthy the southern European countries were decades ago before there was wealth and a cultural change.

The debris, desolation and filth are in developing countries that don't have cheap coal-fired electricity. In the Western world, the elder-

---

39: *Laudato Si'*, Paragraph 161
40: *Laudato Si'*, Paragraph 21
41: Bailey, Ronald 2015: *The end of doom*. Thomas Dunne

ly certainly do lament that once beautiful landscapes are now covered with rubbish. These are wind industrial complexes, which the Pope promotes.[42] What we are seeing today, especially in the developing countries, is a migration of people from rural areas to the cities. This is good news for the forests, wildlife and biodiversity. Oh … and by the way, we have not run out of land, food, water or energy and mineral resources[43] as has been predicted for hundreds of years.

The Pope claims:

> *The earth's resources are also being plundered because of short-sighted approaches to the economy, commerce and production. The loss of forests and woodlands entails the loss of species …*[44]

This is claptrap. The Pope's advisors should consult the World Bank or the UN who show that resource use is based on supply and demand, resources are replaced or recycled, forest areas of the planet are increasing and there is normal species turnover with no evidence for a significant human-induced extinction (sometimes emotionally and erroneously called the Sixth Mass Extinction). To escape from poverty and with population increase, more resources are needed. The choice is simple: less poverty due to resource consumption (especially coal) with resource husbandry or condemnation of poor people to eternal poverty.

Western nations are indeed fortunate to have inherited from the Enlightenment a system of democratic governments that have used science, engineering and logic to address the problems of society. We see many wonderful 19th century structures still standing proudly today. The advances of science in the 19th and 20th centuries gave us mechanisms of evolution and genetic inheritance; the understanding, splitting and harnessing of the atoms; micro-electronics and cheap

---

42: *Laudato Si'*, Paragraph 228
43: *Laudato Si'*, Paragraph 27
44: *Laudato Si'*, Paragraph 32

rapid communication systems; and an explanation and an integration of palaeoclimates and continental drift.

During difficult times such as World War II, polio epidemics and food shortages, governments relied on science for survival. In modern comfortable times, governments have become increasingly divorced from science, which has been captured by green left environmental activists and party politics. Science is now not seen as a mechanism of fulfilling curiosity, understanding and solving problems for society. In some quarters, it is seen as a process that gets in the way of ideology. At present, the greatest threat to the Western world is not the despoiling of the environment. It is self-loathing. Many now no longer are prepared to defend the prize that took 2,000 years to create.

The Pope shows concern about societies despoiling the environment by devegetation[45] yet the Industrial Revolution shows that when societies switched to cheap and reliable coal-fired electricity, the environment improved. Balance and perspective were needed in the papal Encyclical rather than parroting green propaganda. And what has all this burning of coal done to the Earth's atmosphere since the Industrial Revolution? Some 400 Gigatonnes of CO_2 have entered the atmosphere. This is a monstrously large number; we should be ashamed of ourselves and, as a result of our guilt, we should give ourselves a thorough thrashing. But should we?

The total amount of CO_2 in the ocean-atmosphere system is 32,000 Gigatonnes and there is at least two orders of magnitude more CO_2 locked up in sedimentary rocks, metamorphic rocks and igneous rocks (especially in the Earth's mantle). Furthermore, even though 400 Gigatonnes of CO_2 have been released by human activities over the last 250 years (1.25% of the ocean-atmosphere system), atmospheric CO_2 has a residency time in the atmosphere of about

45: *Laudato Si'*, Paragraph 32

five years before natural sequestration into plants, marine life, ocean water and sediments.

However, the Pope objects to the use of coal.[46] The burning of coal in the Industrial Revolution gave steam power, which evolved into the internal combustion engine and national electricity grids, which now are used for employment-generating industries and domestic comforts (central heating, air conditioning, cooking, hot water, entertainment and communications). History has shown us that contemporary pollution is at its worst in the intermediate stages of the growth of developing economies and, once societies become wealthier, some of the wealth is used to tackle pollution and environmental problems.[47] We are currently seeing China and India in the intermediate stages of development. Both have massive pollution and both countries are now starting to divert funds to tackle environmental and pollution problems derived from rapid industrialisation.

Some in the Western world feel that they have a moral obligation to help developing countries like India lift their people out of poverty. This can easily be done by building coal-fired power stations. And this is exactly what is happening. The Pope is certainly correct that we "*may well be leaving to coming generations debris, desolation and filth*"[48] but he does not state the caveat: Only if there is reduced economic growth.

Before the 20th century, the big killers were infections and viral diseases. One scratch in the farmyard could mean tetanus or blood poisoning. One infected tooth could be a death sentence. Childbirth was a lottery where the Grim Reaper had high odds. Pneumonia gave gravediggers busy winters.

46: *Laudato Si'*, Paragraph 165
47: Brimblecombe, P. 1987: *The Big Smoke: A history of air pollution in London since Medieval times*. Routledge, Kegan and Paul
48: *Laudato Si'*, Paragraph 161

What if Greenpeace existed at the time of the Industrial Revolution and stopped the use of coal?

Romantic times

The Enlightenment was followed by an opposing intellectual movement that arose in the late 18th century and persisted to the mid 19th century. This was the Romantic era, which was partly in response to the Industrial Revolution. Liberalism, radicalism and nationalism grew and Romanticism preferred emotion and aesthetics to the rationalism of the Enlightenment. Heroic individuals were highly valued as were art, literature, music, nature, education and the natural sciences. There was a misty-eyed nostalgic view of medieval times and fairs and events replete with medieval costumes were common.

In the English-speaking world, poetry thrived (e.g. William Wordsworth, Samuel Taylor Coleridge, John Keats, Lord Byron, Percy Bysshe Shelley) and literature across the UK, the continent and USA captured the imagination of readers (e.g. Sir Walter Scott, Robert Burns, Thomas Moore, Jane Austen, Bronte sisters, François-René de Chateaubriand, Victor Hugo, Alexandre Dumas, Goethe, Pushkin, Lermontov, Washington Irving, Walt Whitman). Landscape painting (e.g. Constable, Turner) became widespread and music was more restricted to Germanic romanticism (e.g. Mozart, Haydn, Beethoven, Schumann, Schubert, Liszt, Wagner, Mendelssohn) with great works from France (e.g. Berlioz) and Italy (e.g. Verdi).

Another Enlightenment

My grandparents were born at the end of the Romantic era. Besides experiencing two world wars and a depression, my Anglospheric grandparents saw the invention of radio, television, radar, photography, phonographs, telephones and other communication technologies, reticulated electricity, motor cars and cycles, rockets, aeroplanes, space travel, nuclear power and the experimentation with now failed socialist regimes (e.g. USSR, eastern Europe, Cuba). Their lives were far better than those of their grandparents.

The plough stayed behind the shed, the horses were pensioned off and broad acre farming was done using tractors. Advances in chemistry meant that synthetic nitrogen-bearing fertilisers could be made from air (and using energy from coal), international trade allowed the importation of potash fertilisers and superphosphate and advances in breeding gave greater plant and animal yields. Many backbreaking jobs were done by machines and new skills needed to be acquired to operate machines. It was the thinking from the Industrial Revolution, science and technology that allowed greater yields of food, fibre and flesh. Various chemicals were invented to mimic natural chemicals to kills bugs and fungi and, as a result, crop losses to disease were reduced. Steel wire made it possible to fence large paddocks for farming and grazing over large areas. Shepherds lost their jobs.

The 1950s was a period of great hope, every second young kid wanted to become a scientist or engineer and great national projects were possible without the blight of green left environmental activism (e.g. Snowy Mountains Hydroelectric Scheme). It would not be possible to build the Sydney Harbour Bridge or the Snowy Mountains Hydroelectric Scheme under the dark cloud of today's green left environmentalism. I'm sure that there would have been a previously unknown left-footed Morris dancing lizard endemic to the site of one of the Sydney Harbour Bridge pylons.

This was another great period of Enlightenment and, although my grandparents had frugal lives, the world was becoming a far better place in their lives. Unlike their grandparents, there was no fear of hunger yet the cold carried away the elderly. There was certainly no fear of warmer weather. There was hope for a better world and the world became a better place.

Yawn, romanticism again

We are going through another period of romanticism in the Western world when former times with alleged social and environmental harmony are seen with nostalgia. Today is the very best time to live on planet Earth. There were no noble savages, it is not romantic to be a beast of burden and it is certainly not romantic to be without mod cons. In the Western world, it is really an age of entitlement where, without bending the back, many think they are entitled to all sorts of benefits. The wealthy Western world is a me-me-me culture, especially among younger people. This was noted by the Pope in his Encyclical.[49]

We now have the comfort of uninformed opinion without the discomfort of thought. Many of the great advances of the Enlightenment have been lost. Universities and institutes are now places of group-think rather than vibrant places with competing ideas. In the 18th-19th century Enlightenment, science, universities and scientific societies were privately funded. Romanticism in the modern world has twinges of the Inquisition whereby those with a different opinion on political matters such as human-induced global warming, same sex marriage or other social matters get attacked in confected outrage on social media if they don't conform to the populist group-think. No cogent arguments are presented, no compromise is sought, no attempt is made to understand an opposing position and the style of discourse is *ad hominem* attack.

In the modern enlightenment, science, universities and scientific societies are funded by governments. He who pays the piper calls the tune. There is no great art, poetry, literature, music or philosophy evolving out of modern romanticism. Education has no value, the art of conversation and deep thought have been lost and politics is no longer a calling to make the world a better place. Modern roman-

---

49: *Laudato Si'*, Paragraphs 26, 149, 204 and 230

tics gain all the benefits of a modern industrial society with better
health, food, security and communications yet yearn for a simple
world without industry. This is not the first time in the history of
man that society has had irrational fads unrelated to history and reality. All previous generations feared the cold times. Cold weather
killed people. It still does. Modern romantics now fear the potential
of warm times.

The depth to which modern romanticism plumbs is exemplified
by Facebook photographs of the latest meal, pet trick or child vomit
(and the dreadful shallow narcissistic suffering that follows); excite-
ment is clicking like, making a comment or sharing on Facebook; intellectual life comprises 140 characters of political wisdom on Twitter and banal electronic petitions on complex matters with just yes
or no answers. The vulgarity, abuse and intolerance on social media
suggest that society has taken a great backward step.

Social media allows people to object to a proposed new coal
mine without ever having been to a mine, worked in a mine, living
anywhere near the proposed mine or knowing how basics such as
electricity, water, food and money are created. The Romantic period
arose in the 18th-19th centuries as an opposing intellectual movement
to the Industrial Revolution whereas the modern romanticism has
arisen as an opposing anti-intellectual force against the recent Enlightenment and the current Industrial Revolution that is taking place
in China, East Asia and India.

People have a great ability to dupe themselves, especially if the
word socialist is used. Many people think that without cheap coal-
fired electricity, the world would be a better place. I remember the
frequent blackouts as a child. There was a big program to build
coal-fired power stations and the Snowy Mountains Hydroelectric
Scheme. Once, we had reliable cheap electricity, there was enough
energy to pump water for flushing toilets, to run fridges and to cook.
The modern environmental romantics should do themselves a fa-

vour and live in idyllic subsistence in a Third World country with no cheap coal-fired electricity. The modern romantics don't understand that without coal they could not live the quality of life to which they think they are entitled. When I see the caves, desert islands and isolated parts of the outback inhabited by romantics living off the land, then I'll stop describing them as hypocrites.

The wealthy Anglosphere

The Anglosphere has evolved from a 2,000-year history. It has the law, tradition, property rights, individual liberties, science, engineering and modern agriculture that extend from Scotland to New Zealand and includes the largest multicultural democracies in the world (USA and India). During this long road from ancient Greek to modern times, we got better at science, engineering, mathematics and medicine.

As a result, we can build immense stable structures; explore the deep oceans, Moon, Mars, Pluto and large asteroids; travel very fast; and communicate instantly. This journey of more than 2,000 years has given us longevity, better health, more security, more wealth and a higher standard of living. During this long history, there have also been backward steps. We may be going backwards now.

The Western world is wealthier because it is more developed and is less corrupt than other parts of the world. This development did not happen by accident. It has occurred because the resources of the Earth have been united with the capacities of human brains and the great institutions of society (which include the Catholic Church). In my career as a geologist I have visited many tragically poor Third World countries with untapped human, mineral and agricultural assets with poverty exacerbated by despotic or corrupt political systems.

Nowhere in the Encyclical does the Pope comment on political systems that create poverty, slavery and premature death although communism gets a guernsey regarding its use of technology to kill people.[50] However, in many places the Encyclical refers to corruption in both poor and wealthy countries[51] but does not refer to the endemic corruption in Third World countries as a barrier stopping people escaping from poverty.

Wealth is a solution to real and perceived environmental problems. Wealthy societies have longevity, a lower birth rate, more and better food, cheap energy hence less necessity to chop down forests or overgraze and funds to solve environmental issues. The average African does not care about the climate, they are worried about where the next meal comes from and, in the longer term, survival, health care, clean water, light, heating, a roof over their heads and education for their children. There is no forthcoming climate disaster in 100 years time for the average African, it is an hour-by-hour struggle for survival. In the developing world, there are billions of people in dire poverty suffering all the ills it brings such as malnutrition, preventable disease and premature death.

The Pope is in effect asking the developing world to ignore the cheapest known available sources of energy and asking them to delay the conquest of malnutrition and to perpetuate the incidence of preventable disease. Denying supply of coal the Third World means that there is no cheap energy, forests are cut down and mothers and their children die unnecessarily. No solar or wind technologies will solve these problems because there is not the money to subsidise ideological unreliable electricity. With many developments in the Third World, the best thing poor people can do is to steal copper wiring, steel and solar cells and sell for the next meal.

The Pope's energy aspirations not only affect the Third World. In the UK, the most blatant transfer of wealth from the poor to the

---

50: *Laudato Si'*, Paragraph 104
51: *Laudato Si'*, Paragraphs 55,123, 142, 172, 179, 182 and 197

rich has been devised around renewable energy. This is supported by the Pope and the left who, *inter alia*, argue that the Church cares for the poor. In the UK, heavily subsidised wealthy landlords have wind turbines on their land so that the poor can be supplied with one of the most expensive, unreliable forms of electricity ever created.

In the Anglosphere, people who work for a living are now outnumbered by those who vote for a living. Political power is now based in cities and those in rural and outback areas of Australia producing food and commodities necessary for a vibrant economy are outvoted by those totally disconnected from reality.

2

WHO'S WHO IN THE ZOO

We all want a clean planet for ourselves and future generations. However, when it comes to matters of green left environmental activism we need to be very wary of idealism, self-interest and a lust for power. This was well enunciated at Bertrand's Russell's 11 December 1950 Nobel Prize[52] acceptance speech on "What desires are politically important":

And among those occasions on which people fall below self-interest are most of the occasions on which they are convinced that they are acting from idealistic motives. Much that passes as idealism is disguised hatred or disguised love of power. When you see large masses of men swayed by what appear to be noble motives, it is as well to look below the surface and ask yourself what it is that makes these motives effective. It is partly because it is so easy to be taken in by a façade of nobility ...

WHO TOLD THE POPE WHAT?

What the Pope was not told

It has not been shown that human emissions of CO_2 drive global warming. It has been long known by those empirical scientists who have no measured evidence for CO_2 driving climate change that climate (and hence air temperatures) are driven in the very long term by the position of the continents and 143-million year (Ma) long galactic cycles (as shown by hundreds of millions of years of sedi-

---

52: Frenz, Horst 1969: *Nobel Lectures, Literature 1901-1967*. Elsevier

mentation in Central Europe),[53] in the not so long term by Milank-
ovitch Cycles (orbital 100,000-year,[54] 41,000[55] and 21,000 years[56]), in
the medium term by solar (1500-, 217-, 87- and 22-year cycles) and
ocean cycles (Atlantic Multidecadal Oscillation, Pacific Decadal Oscillation) and in the short term by ocean and atmosphere cycles
(Lunar tidal nodes, El Niño-Southern Oscillation). Was the Pope
told about the Earth's climate cycles?

Filling of the atmosphere with aerosols from volcanic eruptions
induces short term cooling such as the five years of cool climate
after Krakatoa (1883) and Tarawera (1886). Asteroids do the same
but large asteroidal impacts are catastrophic events where one has to
quickly kiss body parts goodbye. In the scheme of things, it is not
worth getting into a lather about a modelled temperature increase
of 0.1°C, this is the temperature change that occurs when you move
your head. A complex climate system having been reduced to the
impact of one variable (human emissions of CO_2) shows the intellectual vacuosity of the green left environmental activists. The Encyclical shows that this is the advice the Pope received.

The Pope writes:[57]

> ... the Church does not presume to settle scientific questions or to
> replace politics. But I am concerned to encourage an honest and open
> debate so that particular interests or ideologies will not prejudice the
> common good.

There has never been an honest debate on environmental and
climate matters. It appears that the Pope's advisors have also not
been honest. The Pope should have been informed by his advisors

---

53: Brink, H-J. 2014: Singnale der Milchstraße verborgen in der Sedimentfüllung
Zentraleuropäischen Beckensystems? Z. Dt. Ges. Geowiss. 166: 9-20
54: Orbital eccentricity
55: Axial tilt or obliquity
56: The combined effect of two precessions
57: Laudato Si', Paragraph 188

that the climate "debate" is not about science. It is about politics. He has joined the haters at the expense of the debaters. The Pope does not seem to be aware that, unlike much science, the human-induced global warming theory has no practical value but has tremendous political value. The Pope should have been told that the predictions of climate "scientists" have failed. Not just one or two of them but all of them have failed. My long life in science makes one wary of predictions. In the Pope's lifetime, no one could have predicted the rise and fall of communism, the rise of India and China, the invention of the internet and the ease of international travel. Who in the Vatican was sceptical of predictions of dramatic climate changes in a century?

The Pope should have been advised that the IPCC very selectively used the scientific literature, used unpublished and unrefereed works and seems to ignore key scientific works on the Sun, clouds, past climate changes and the geological past when the atmospheric CO_2 content was far higher than at present. Who did the background checks on the IPCC? The IPCC is a deeply flawed lobby group mired in scandal. The Pope ignores the long history of planet Earth. Nowhere in the Encyclical does he refer to the uncertainty of science. Was he ever told that there have been more than 18 years with no warming? The Encyclical does not discuss the fact that there is no relationship between atmospheric CO_2 and climate over time.

There is no mention of the blatant fraud of the 2009 Climategate emails, the Michael Mann fabrication of a hockey stick and the worldwide "adjustment" of the temperature record. The Vatican must have records to show that the Dark Ages, Medieval Warming and Little Ice Age existed yet has allied itself with people who want to change history and expunge such natural climate variations from the record. The Pope is concerned about education[58] yet was not informed that green left environmental activists groups like Greenpeace and WWF have worked relentlessly for decades in the edu-

---

58: *Laudato Si'*, Paragraph 94

cation system to sow misinformation about human-induced global warming. The Pope was very clearly told that there is a scientific consensus on human-induced climate change[59] yet I show in this book, as many others have,[60] that this is far from the truth.

The Encyclical contains nothing on the economic, social, environmental and human medical costs of renewable energy. Green left environmental activists promote renewable energy with no understanding of costs and environmental damage. The Pope used the word "profit" 15 times in a pejorative sense and does not seem to understand that there is a queue of greedy bankers lined up behind the renewables industry.

If the Pope is suspicious of organisations that make a profit then the alarm bells should have rung very loudly about the renewables industry. If the Pope had been informed of how much climate "science" and the renewables energy industry are supported by taxpayers, he might have had something different to say about profit. If just a fraction of the money spent on a disproven theory went to infrastructure in the developing world, then there would be far fewer poor people for the Pope to be concerned about.

Who failed to tell the Pope the long and short climate cycles are unrelated to consumption, the modern world or human emissions of CO_2, that atmospheric CO_2 increases after a temperature rise not before and hence cannot drive warming and that the polar ice caps, sea ice and glaciers come and go for a diversity of reasons? He is promoting post-modernist science where it is alleged the facts are certain, contrary facts are ignored, the stakes are high and decisions are urgent.

The Pope has presented the very narrow view of green left environmental activists and not thought through the implications of his

---

59: *Laudato Si'*, Paragraph 23
60: Laframboise, Donna 2011: *The delinquent teenager who was mistaken for the world's top climate expert*. Connor Court

Encyclical. If followed to the logical conclusion, the Pope's Encyclical on climate and the environment will keep the poor in eternal poverty.

Why was the Pope so poorly advised? Who were the advisors? What was the agenda of the advisors? It is clear that the Pope is on the road to Paris but without a map.

Birds of a feather

As an Argentinian Cardinal, the Pope was struck by floods and unsanitary conditions in shanty towns in his home country known as misery villages. The Pope certainly would have seen filth, lack of care for workers and the brutality of emerging capitalism in Argentina. This is not the capitalism of the modern Western world. One size does not fit all. In January 2015 after a visit to the Philippines, the Pope announced that he wanted to make a contribution before the UN's conference on climate change in December 2015 in Paris. It's clear that the conclusions of the Encyclical were set when the Pope was an Argentinian Cardinal and the Pope signalled his intent to produce a major document on the environment soon after being elected in March 2013. It is no surprise that the Pope is now facing opposition from conservative theologians.

A member of the Club of Rome, the self-professed atheist and radical environmentalist Hans Joachim Schellnhuber, had a part in drafting the papal Encyclical and was apparently selected for this role by Archbishop Marcelo Sanchez Sorondo, the head of the Pontifical Academy of Sciences. It is no wonder that the Encyclical reads like a Club of Rome or Greenpeace doomsday disaster document. The Pope, untrained in natural sciences, needs advice and it is clear that the environmental "experts" provided the answer the Pontifical Academy of Sciences desired.

The litany of simple scientific errors, the scientific dogmatism and omissions of simple science suggests that the Pontifical Academy of Sciences was not interested in discussing science and finding clever solutions that benefit humans while protecting the planet. The anti-science, anti-technology and anti-industry agenda pushed in the Encyclical[61] is that of green left environmental alarmists wanting to push us back in the caves. There might be enough caves on Earth for seven thousand people but not for more than seven billion people.

Naomi Klein and Cardinal Peter Turkson led a high level conference on the environment in Rome in late April 2015. The conference was sponsored by the Pontifical Academy of Sciences and attended by the UN Secretary-General Ban Ki-Moon. Klein and Turkson were joined by men of the cloth and green left environmental activists. Scientists with a different opinion were not allowed to attend, let alone be heard. If Klein's book[62] is any guide, then she is not a fit and proper person to advise the Pope. In her book she claims:

> *Carbon dioxide stays in the atmosphere one to two centuries with some of it remaining for a millennium or more.*

Wrong. The only time this happened in the history of the planet was during the first few hundred million years when there was no water on the surface of the Earth. The atmospheric retention time for CO_2 in the atmosphere is about 5 years. This has been shown from the 116 above-ground Russian nuclear tests that gen-

---

61: *Laudato Si'*, Paragraphs 9, 16, 20, 54, 60, 102, 106, 107, 108, 109, 110, 112, 114, 131, 132, 136, 165 and 172

62: Klein, Naomi 2014: *This changes everything: Capitalism vs. The climate*. Simon and Schuster

erated short-lived carbon 14 $(C^{14})^{63}$ and is inconsistent with IPCC estimates.[64]

She also writes that the "Medieval Warm Period was thoroughly debunked long ago."

Wrong. In fact, there is validation from a diversity of disciplines, the measurement of a diversity of climate proxies[65,66,67] is in accord with the recorded history[68] (i.e. the coherence criterion of science) and to rewrite history to suit ideology gives a window into Klein's agenda and intellectual dishonesty. Time and time again studies from all four corners of the globe show that the Medieval Warming was global.[69,70] According to Klein, if a fact contradicts ideology, then get

---

63: There are three natural isotopes of carbon. C^{12} (98.9%) and C^{13} (1.1%) are stable isotopes, these ratios can vary slightly due to fractionation by thermodynamic, biological and geological processes. C^{14} is radiogenic from cosmic radiation (and atomic bombs) and has a half-life of 5,730 years. This means that after 5,730 years a bucket full of C^{14} would have decayed to half a bucket of C^{14}, after another 5,730 years of decay a quarter of a bucket of C^{14} would remain, after another 5,730 years and eighth of a bucket of C^{14} would remain and so on.

64: http://wattsupwiththat.com/2013/07/01/the-bombtest-curve-and-its-implication-for-atmospheric-carbon-dioxide-residency-time/

65: Broecker, W. S. 2001: Was the Medieval Warm Period global? *Science* 291: 1497-1499

66: Huang, S. *et al.* 1997: Late Quaternary temperature change seen in worldwide continental heat flow measurements. *Geophys. Res. Lett.* 24: 1947-1950

67: Huffman, T. N. 1996: Archaeological evidence for climate change during the last 2,000 years in southern Africa. *Quat. Internat.* 33: 55-60

68: Fagan, Brian 1999: *Floods, famines and emperors: El Niño and the fate of civilizations.* Basic Books

69: Ouellet-Bernier M.-M. *et al.* 2014: Paleoceanographic changes in the Disko Bugt area, West Greenland, during the Holocene. *The Holocene* 24: 1573-1583

70: Yan, H. *et al.* 2015: A composite sea surface temperature record of the northern South China Sea for the past 2,500 years: A unique look into seasonality and seasonal climate changes during warm and cold periods. *Earth Sci. Rev.* 141: 122-135

rid of the fact and cling on to the ideology. Just because a conclusion is uncomfortable does not mean is can be dismissed with a throw away line.

The Klein book is not about climate. It is about green left environmental socialism:

> *Lowering emissions is just one example of how the climate emergency could – by virtue of its urgency and that the fact that it impacts virtually everyone on earth – breathe new life into a political goal ... (such as) raising taxes on the rich, blocking harmful trade deals, reinvesting in the public sphere ...*

and

> *Even more importantly, the climate movement offers an overarching narrative in which everything from the fight for good jobs to justice for immigrants, reparations for the injustice of slavery ... all becomes a grand project of building a non-toxic shock-proof economy before it is too late.*

So that's it. The science is not important and we appear to have an unmeasured climate emergency. What really matters is what social "justice" can be squeezed from climate change. This is a book by a non-scientific social activist. Some of us may agree with some of her social justice causes but fear-mongering scientific fraud is not the way to achieve a result. Naomi Klein, a ferocious critic of 21st century capitalism, now views the Pope as an environmental campaigner. This is the quality of person that the Pope uses for advice. No wonder the Encyclical has been attacked widely. Some of the language (e.g. use of the word urgent[71]) has it reading like one of the alarmist documents of a green left environmental activist group such as Greenpeace. Is this the fingerprint of Naomi Klein?

71: *Laudato Si'*, Paragraphs 13, 14, 26, 31, 57, 111, 114, 141, 162, 173, 175, 181, 189 and 192

A multi-faith march through the Eternal City preceded the Vatican welcome to green left environmental campaigners (including Greenpeace, Oxfam and self-professed atheists). The conference focused on the UN's impending climate change summit in Paris. The Gaia-worshipping Greenpeace activist, Kert Davies,[72] is celebrating the Pope's entry into the climate debate.

The presentation of the papal Encyclical was by a panel of five including a Rome school teacher and Hans Joachim Schellnhuber, who used his 15 seconds of glory to give a lesson about climate "science." The man who had the ear of the Pope stated:[73]

> *It is the consumption of the upper and middle classes that is destabilising the climate, and it's not the quest of the poor to have some access to resources, like clean water. It's not poverty that's destroying the planet, it's wealth.*

I'm sure the readers felt guilty, bought more indulgences and didn't analyse this statement. The argument I put in this book is that the rise of the middle class in the first Industrial Revolution and the increase in the middle class in the current Industrial Revolution is saving hundreds of millions of people from poverty. Only wealth can stop poverty.

Schellnhuber is Chairman of the German Advisory Council on Global Change. He has a master plan for society, albeit on the authoritarian side:

> *The German Advisory Council on Global Change, which I chair, will soon unveil a master plan for the transformation of society.*

Those of us who have a sense of history felt our hair stand on end when we hear the words master plan coming out of Germany

---

72: http://www.climatedepot.com/2015/02/26/regurgitate-unsupportable-accusations-greenpeaces-kert-davies-is-back-again/

73: *Socialist Moaning Herald*, 18 June 2015 (sometimes called the *Inner Sydney Moaning Herald*)

and being used in the context of changing the political landscape of the world. Schellnhuber further states:

> *The German Advisory Council on Global Change recommends reducing CO_2 emissions from fossil fuels to zero by 2070 at the latest. This policy is both ambitious and incisive, because the zero target must be reached by every country, every municipality, every company and every citizen if the world as a whole is to become climate-neutral.*

Are we all to stop breathing? We inhale 0.04% CO_2 and exhale about 4%. There was no practical method suggested as to how there would be no human emissions of CO_2 by 2070 and, if it's OK by you Dr Schellnhuber, I want to get on with my life without having to face some sort of global German authoritarianism. Does Schellnhuber realise that if he is successful, in 55 years time his world will be cold, poor and have the health and longevity we had 300 years ago. In a *Nature* article[74] in 1999, he stated:

> *Although effects such as glaciations may still be interpreted as over-reactions to small disturbances – a kind of cathartic geophysiological fever – the main events resulting in accelerated maturation by shock treatment, indicate that Gaia faces a powerful antagonist.*

I'm sure we all know what he meant! This is post-modernist gobbledygook. The embrace of Gaia by an atheist should have sent the Pope's advisors scurrying to look for a more balanced and hopefully Christian climate expert. There were some waiting in Rome at the time of the April 2015 conference.

Was the Pope or his advisors fearful of hearing another view?

74: Schellnhuber, H. J. 1999: 'Earth system' analysis and the second Copernican revolution. *Nature* 402 (Supp.) C19-C23

A contingent of experts[75] went to Rome to express an alternative view; they were kept waiting and received neither an audience with the Pope nor his advisors. Philippe de Larminat,[76] who authored a book on climate change, was invited in March 2015 by Cardinal Peter Turkson to the Vatican for the April 2015 meeting and, five days before the meeting, was uninvited. He had already booked his flight to Rome.[77]

Was this because de Larminat wrote that solar activity is the main driver of climate change rather than CO_2 emissions by humans? In April 2015 the Pope had clearly already decided that climate change was man-made. The Encyclical was released in mid-2015 claiming that climate change was man-made. The words in the Encyclical "… forthright and honest debate …",[78] "…open debate…",[79] "…broad, responsible scientific and social debate …",[80] "… full debate …"[81] and "…honest and open debate …"[82] are completely hollow and incommensurate with the fact that only a green left environmental activist view was promoted.

75: Dr E. Calvin Beisner (Cornwall Alliance for the Stewardship of Creation), Hal Doiron (former NASA Skylab and Space Shuttle engineer), Dr Robert Keen (meteorology lecturer at the University of Colorado), Viscount Christopher Monckton (Science and Public Policy Institute and a Catholic), Marc Morano (Heartland Institute, ClimateDepot.com), Dr Tom Sheehan (Science and Environment Policy Project) and Elizabeth Yore, JD (former General Counsel at the National Center for Missing and Exploited Children)

76: Philippe de Larminat, 2014: *Climate change: identification and projections.* Wiley

77 http://.dailymnail.co.uk/news/article-3133468/French-climate-change-doubter-uninvited-Vatican-summit-weeks-Pope-declared-global-warming-man-problem.html

78: *Laudato Si'.* Paragraph 16

79: *Laudato Si',* Paragraph 61

80: *Laudato Si,* Paragraph 135

81: *Laudato Si',* Paragraph 182

82: *Laudato Si',* Paragraph 188

For the Pope to take advice from self-processed atheists demeans the deeply held faith of his flock. The Pope surely must be aware of the quote attributed to G. K. Chesterton (1874-1936): "When a man stops believing in God he doesn't then believe in nothing, he believes anything."[83]

The green left environmental activists will believe anything. In this book, I show that they will even believe in human-induced climate change in the absence of evidence. Furthermore, when presented with contrary evidence, they ignore it and still believe in anything.

The Pope has aligned himself with the worst of the pagan, anti-religious and Gaia-worshipping green left environmental activists who will exploit him and his Encyclical for their own ends. Pro-market economic reformers are painted as sinful, giving a huge moral boost to Marxists. In seeking reconciliation between science and religion, the Pope has firmly come down on the side of anti-scientific political activists.

Embrace of communism

It is generally acknowledged that Saint John Paul II (together with Prime Minister Thatcher of the UK and President Reagan of the USA) were the driving forces responsible for the deconstruction of Eastern European communism. This enriched the world spiritually, morally, environmentally and economically. The threat of a global war was greatly reduced. Pope John Paul II, Thatcher and Reagan stopped the Cold War and made the Earth a more peaceful place. Under Pope John Paul II, the Church had once again become a powerful global force.

Pope John Paul II was the first non-Italian Pope for 400 years. He was born Karol Józef Wojtyla, experienced the Nazi invasion of Poland when he was a young man and then the subjugation of Po-

---

83: http://www.chesterton.org/ceases-to-worship/

land by the communist USSR at the end of World War II. For most of his life he lived under totalitarian political systems. He certainly knew about totalitarianism, communism and socialism dressed up as humanism.

The Pope's 2015 Encyclical is a recipe for catastrophe and such prescriptions can only corrode the credibility of the Church and reverse the gains made by Pope John Paul II. I wonder if the Church's role in the destruction of Soviet and Eastern European communism will be taught in schools as a basic historical truth or whether a green left environmental activist negative view of the Church will be taught?

Pope Francis quotes the first Encyclical of Pope John Paul II[84] yet, a few weeks after the release of *Laudato Si*, the Pope was given a carved wooden communist crucifix from Bolivia's socialist President Morales in July 2015.[85] It was a sculpture of Christ nailed to the communist emblem of a hammer and sickle made by the murdered Jesuit priest Luis Espinal. Pope Francis is the first Jesuit Pope. Bolivian Bishop Gonzalo del Castillo denounced the gift as a "provocation, a joke."[86]

The hammer and sickle crucifix would sicken many, especially in Poland, where Pope John Paul II helped tens of millions of people escape from the iron grip of communism. Those who experienced the tyranny and repression of atheistic communism would have some disquiet about the current Pope embracing a competing despotic atheistic ideology. The Pope returned the crucifix later.

Alarm bells immediately ring loudly because the familiar catch

---

84: *Laudato Si'*, Paragraph 5

85: http://www.catholicherald.co.uk/news/2015/07/09/pope-francis-praises-reforms-of-evo-morales-on-arrival-in-bolivia/

86: http://www.news.com.au/world/south-america/bolivian-president-evo-morales-presents-pope-francis-with-communist-crucifix/story-fnh81jzo-1227437028470

cries of green left environmental activists and communists are used,[87] such as:

A very solid scientific consensus indicates that we are presently witnessing a disturbing warming of the climatic system.

As I show later, there is no consensus. A scientific consensus is oxymoronic. Consensus is a word used in politics, not science. Measurements over the last 18 years don't show a warming let alone a *"disturbing warming"*. There has been a general warming trend over the last 350 years since the Maunder Minimum. This is exactly what would be expected after the coldest period of the Little Ice Age. During this warming trend there were cooling cycles and times with neither cooling nor warming. The Pope's advisors forgot to tell him this vital piece of evidence.

Papal leadership

In the US, a Gallup poll shows that Pope Francis' popularity decreased after publishing his Encyclical.[88] Soon after he was elected, his popularity was 58% (April 2013), it rose to 76% in early 2014 and since publication of the Encyclical it has fallen to 59%. In 2014, 89% of US Catholics had a favourable view of the Pope. It is now 71%. In 2014, 72% of US Conservative voters had a favourable view of the Pope. This has now dropped to 45%. In the US, Pope Francis rated higher than Pope Benedict XVI and lower than Pope John Paul II.

The Encyclical is not a document of a leader. It follows simplistic Western populist green left environmental activist thought and promotes the idea that if young people demand change, then it should happen.[89] Maybe the Pope seeks celebrity status and wants to hitch his horse to the environmental bandwagon so strongly sup-

---

87: *Laudato Si'*, Paragraph 23
88: http://gallup.com/poll/184283/pope-francis-favorable-rating-drops.aspx
89: *"Young people demand change."* *Laudato Si'*, Paragraph 13

ported by the young. If this is his leadership, then he should have been reminded of the quote attributed to the 19th century politician Alexandre Auguste Ledru-Rollin, head of state in France for 54 days in 1848:

There go the people, I must follow them for I am their leader.

Global disruption constantly occurs. It is unsettling for workers, companies and governments. It is a reality that the Pope cannot accept[90] and, as a leader in a constantly changing world, he shows great fear of technology concurrently with lauding some technological achievements.[91] It seems that the Pope longs for the static world of long ago. This was the world of poverty, disease and despots where people died like flies. The rate of change of technology is very fast, most of us are struggling to keep up and creative technology is changing the world. It is the latest technology that allows the papal Encyclical to be published and circulated, it is the latest technology that allows the Pope to visit his flock in South America and it is the latest technology that churns out CO_2 to allow papal advisors to come to Rome and inform his Holiness of the dangers of CO_2.

Constant change, brought on by technological innovation and instant worldwide capital flow, produces economic forces that reshape the meaning of work, reshape jobs, reshape industries and change nations. Pity help a nation that does not change to dynamic economic forces by being inflexible and having Luddite views about technology. No nation is immune from such forces, especially if it is running at a deficit. Yet many Western nations are dominated by poor leadership, failed ideas, false debates and trivial issues. Capitalism is creative, technology is liberating people from factory production lines and new jobs are created (e.g. digital revolution). We should be celebrating. The Pope does not.

90: *Laudato Si'*, Paragraph 51
91: *Laudato Si'*, Paragraphs 9, 16, 20, 54, 60, 102, 106, 107, 108, 109, 110, 112, 114, 131, 132, 136, 165 and 172

How do we know what we know?

The fatal assumption

If we are to have an opinion on climate change, then we require a few fundamentals. We need to have the thinking processes that have evolved over the last 2,000 years.

The entire trillion-dollar climate change industry rests on a single hypothetical assumption. The assumption is that emissions of CO_2 by humans drive global warming. To this day there is still no scientific evidence to support this assumption and, after being berated for three decades about how the world will warm from human emissions of CO_2, some of us have had the heretical thought that maybe the assumption is wrong. Maybe the models are wrong? Maybe the Earth is doing what the Earth does and we humans are pretty insignificant?

Who pays the piper?

I have doubts that CO_2 emissions from human activity drive climate change. The green left environmental activists would claim that this is because we doubters are paid by the fossil fuel industry to espouse such views. This is a window into how the green left environmental activists think and operate with their own venal motives.

There is an increasing amount of evidence to show that many green movements are well funded from a diversity of sources (e.g. unions) and, to use the green left thinking, their motives are highly questionable. Union superannuation funds have huge investments in the renewable energy industry. In Australia, the biggest single donation ever made to a political party was to the Green Party. The next largest donations are from the unions to both the Australian Labor Party and the Green Party.

The public records of political donations show that those evil multibillionaires who apparently run the country, take risks and employ millions of people, are not in the league of union donations and often make donations to both the right and left side of politics. Climate "science" funding is bespoke science for governments. It is not driven by curiosity. It is not driven in an attempt to make the world a better place. We cannot forget President Eisenhower's farewell speech on 17 January 1961 where he warned:

> *The prospect of domination of the nations' scholars by Federal employment, project allocation, and the power of money ... that public policy could itself become the captive of a scientific-technological elite.*

Ike warned us and we let it happen. Political policy has now been captured by a green left environmental activist elite who are driven by money, malice, power and fame. The fear of human-induced climate change is a Heaven-sent opportunity for the green left environmental activists to vastly increase government control over the economy and the personal lives of taxpaying citizens. They have introduced unscientific terms such as *"the science is settled"* and have called those in a pluralist democratic society who do not agree as *"climate change deniers"* (with the intended obvious Nazi Holocaust implications). Settled science is an oxymoron. Science changes with new evidence. An *ad hominem* attack only shows that there is no evidence to support an argument.

I find it very difficult to understand how the science could ever be settled with climate. There are a very large number of inputs and outputs that we don't understand, there is contradictory evidence, the past is not in accord with some theories about the present, the unknown always presents us with surprises and there is always more to learn because science is a veritable cornucopia of unanswered questions. The obvious ones are why did the warming from 1978 to 1998 stop? How sensitive is climate to increased atmospheric CO_2 levels, especially in the light of past times when the atmospheric CO_2 content was very high?

The US government spends $US 2.5 billion per year on research that focuses on CO_2 and ignores the powerful natural forces that always have and always will drive climate change. This research generates papers, books, reports, press releases, models and careers in government-supported science and in the media. Massive egos are fed and instant experts are created *ex nihilo*. We are constantly warned of record high temperatures, more droughts, melting ice caps, sea level rise, more frequent and larger storms, extinctions and all sorts of unprecedented crises. The list of potential disasters is endless.

What has the $US 2.5 billion *per annum* actually achieved? Is the world a better place? How many lives have been saved? How many African villages have been connected to the electricity grid? How much potable water has been produced for the desert people of North Africa? Has all this research made one iota of difference to the global climate? Draconian reductions in CO_2 emissions by the developed world maybe will prevent 0.0001°C of warming, if indeed human emissions of CO_2 actually create global warming. Such a decrease in temperature cannot even be measured. What's the point?

The climate researchers are not happy, they want more money such that they can frighten us more about the environment and glaring anomalies strongly suggest that green left environmental activists just don't care about the environment. For example, wind turbines slice and dice millions of birds and bats each year but are exempt from the US *Endangered Species Act*'s fines and penalties. Imagine if a mine, factory, coal-fired power station or nuclear power station killed the same number of birds and bats and created health problems in humans that lived nearby. They would be shut down immediately, fined and probably sent bankrupt.

The climate business is a huge interlocking self-perpetuating business hoovering up public money. Business just can't believe its luck. This is a transfer of money from the poor to the wealthy. From 2007 to 2013 the corn ethanol interests in the US spent $158 mil-

lion lobbying for more green mandates and subsidies. They provided
$6 million in election campaign subsidies for alcohol-based fuel for
cars that reduces efficiency, damages engines, emits more CO_2 than
conventional hydrocarbons and requires enormous amounts of land,
water and fertilisers.

Universities and independent thinking

Avalanches of taxpayer's funds have been thrown at climate "scientists" to try to show that human emissions of CO_2 drive global warming yet this has not been shown. I have doubts about human-induced
climate change because I have been formally scientifically trained
and have undertaken scientific research for decades. My doubts are
because the science is bad. Really bad. I am not influenced by what
professional societies, politicians, policy makers, media groups, rent
seekers, activists, ideologists and dreamers might think because their
views don't get tainted by evidence. For one side of a quasi debate it's
about the money, for the other side it's about the science.

Modern universities and scientists live off government grants and
are not nearly as independent as we would like to believe. Scientists
who don't abide by dogma are treated as enemies of the state and
many of them only give their considered professional opinion after retirement when they are unshackled from bespoke science. The
profile of the most dangerous person in the current climate claptrap
environment is an older independent polymath who has spent their
life in science, uses critical thinking and analysis, has learnt from mis-
takes, reads history and understands logic. The only way to handle
such people is to besmirch their reputation and pillory them on social media. No counter arguments are ever aired.

I am not influenced by brazen aggressive political-style attacks
questioning contrarian scientists' ethics, morality, competence and
sanity. The attacks are not scientific, they are *ad hominem* and many
in the public have realised that such *ad hominem* attacks are used as

a substitute for argument underpinned by evidence to defend a hypothesis. The one unifying thread amongst we doubters is that we will not deliver what the policy makers want, we will not be told how to think, we are fearlessly independent and we don't believe that models override evidence. We call it as it is. If activists, politicians or the green left environmental activist media don't like it, then bad luck. It doesn't matter who pays us, we can't help ourselves and are annoying nuisances.

We learn from history that humans not only get things wrong but stubbornly hang onto the stupidest ideas right to the bitter end. Just think of a flat Earth or the settled science that showed that the Sun and all planets rotate around the Earth. These ideas took thousands of years to bury.

It continues to amaze me that people who are expert in one branch of science don't appear to be able to apply the same rigorous thought to other areas of life. Much science training today, especially at the second rate universities (many of which are rebadged Colleges of Advanced Education or Technical Colleges) is not rigorous and is simply teaching a trade. With such teaching, there is an implicit suggestion that there is a "right" answer. Epistemology does not seem to be very important. What does a degree in environmental science really mean? Does this mean that the graduate has a full understanding of the chemical sciences, the physical sciences, the earth sciences, the life sciences and the mathematical skills to analyse data? I fear not. Very few science graduates have taken courses in the humanities where the data can be so unreliable that there is no answer to a simple question.

The history of science shows that great scientific discoveries are not made by consensus. Great scientific discoveries were made by individuals paddling upstream and these researchers were generally pretty difficult people to deal with.

Oxymoronic journalistic ignorance

Concurrent with a decline in educational standards has been a decline in journalistic standards. Real journalists collect facts. Facts are sacrosanct. Journalism now resorts to on-line trivia, banality and vulgarity. Gone are the days when an intellectual argument could be mounted in the media. Many networks now collect their news feed from Twitter. Only a few journalists have the skills of rational argument. Rather than a rational argument, much of the media commentary has resorted to sensationalism and vulgar undergraduate abuse. The tribal green left environmental activists among them seem to be a very angry, envious inarticulate lot.

By using the internet, journalists are now able to trawl for an opinion that suits their ideology or their current scare story. They don't have to investigate, ask searching questions and validate the facts they have collected. Once a story is published, they move on to the next trite scare story. If an early story is wrong, then there is neither a correction nor an apology because the journalist has moved on to the next great scare. Scientifically illiterate journalists do not have the knowledge, training, scientific skills, research experience or critical thinking ability to analyse a press release from a scientific organisation, a scientific publication or a claim on a blog site.

They view a press release or claim as fact, repeat the information and publish their own uninformed opinion. A couple of them have actually admitted that they are activists[92] hence one wonders whether enough information is given for the reader, listener or viewer to come to an informed decision. Much journalism comes from retired political hacks (generally from the left) and their political opinion is entwined within the safety of political correctness. Political correctness hides ignorance.

The media now has become the mouthpiece for screeching pro-

92: Fran Kelly, ABC Radio National

paganda. Always be suspicious when a journalist interviews another journalist for an opinion. I can make up my own mind on a diversity of issues and I would prefer the media to be providers of sacrosanct facts and not cheer leaders, political hacks and activists. However, this is the reality of what now passes as journalism. Because of the inability to argue, analyse and critically think, journalists close debate before it happens, allow argument to be dominated by polarised hatred and display no pretence of balance. On television and radio, journalists constantly interrupt and talk over their guests who may have a different opinion and allow a barrage of hatred from a hand-picked audience. What is taken as a discussion is really a group-think with journalists agreeing with journalists using Orwellian newspeak (e.g. carbon pollution, consensus). Such tactics are used to hide ignorance and not inform an audience.

The media views science as a popularity contest in which those with the greatest numbers of votes in the consensus election win. One has to be very wary of political consensus decisions as full knowledge of back-room deals is normally not transparent. If there is a hypothesis that human emissions of CO_2 create global warming, then in science only one item of evidence is needed to show that this hypothesis is wrong. Dozens of items of evidence have been listed here and in other writings[93,94] to show that the human-induced global warming hypothesis is wrong.

Dominant consensus views in the past were that the Sun rotated around Earth, that burning material was due to a substance called phlogiston, that health was driven by humours, that leeches could cure most diseases, that malaria was caused by bad air, that earthquakes were an act of God, that heavier-than-air machines could

---

93: Plimer, Ian 2015: The science and politics of climate change. In: *Climate change: The facts* (Ed. Alan Moran). Institute of Public Affairs, 10-25
94: Plimer, Ian 2009: *Heaven and Earth. Global warming: The missing science.* Connor Court

never fly, that the continents were fixed and that duodenal ulcers were caused by stress. All these views were strongly supported by science and authorities at the time and have all been shown to be wrong.

The skewing of the scientific literature is not healthy for society that depends upon scientists and scientific literature for trustworthy advice for wise policy decisions. My advice to young climatologists: Do not write a paper questioning the IPCC, the popular paradigm of disastrous climate change or your host climate institute. You will destroy your career prospects and you'll never get a research grant. The only way to advance in a career of climate "science" is to be deceitful. The one common thread throughout all climate "science" is misleading and deceptive conduct. Many examples are given in this book.

The media is now digging out old documentaries about natural disasters softening us up for the late 2015 Paris climate conference and we are again seeing for the umpteenth time footage of natural processes such as glaciers breaking off into the sea. All the same old scare stories are being warmed up with greater embellishment and the green left activist scientists are making scarier and scarier predictions. Apparently, disaster is just around the corner as it was just before the Copenhagen conference in 2009.[95] The disaster story with Copenhagen was that there were metres of snow, it was bitterly cold and the public was asked to believe that humans drive global warming. Ask me, I was freezing there at my own cost.

THE SCIENTIFIC JOURNEY

The Greeks

The Greek pre-Socratic philosopher Anaxagoras (510 to 428 BC) was interested in eclipses (which he could correctly predict), meteors,

---

95: 2009 UN Climate Change Conference, Copenhagen 7-18 December 2009

rainbows and the Sun. He argued that the Moon reflected sunlight (correct), that it had mountains (correct) and that it was inhabited (well …). His Earth was flat and supported by 'strong' air and disturbances to this air created earthquakes. His greatest contribution was that he suggested that the Sun was a star, which we now know it is. We also know now that there are many types of stars.

Aristarchus of Samos (~310 to 230 BC) advanced the thinking about the Sun and promoted a heliocentric model of the Solar System (i.e. the Sun was the centre of the Solar System). These ideas were rejected by Aristotle (384 to 332 BC) and Ptolemy (90 to 168 AD) who argued for a geocentric model of the Solar System (i.e. the Earth is the centre of the Solar System). Aristotle influenced scholarship until Newtonian classical mechanics appeared in the Enlightenment. The first formal study of logic was by Aristotle. For two thousand years, science was settled and we all knew that the Earth was the centre of the Solar System. Although Copernicus and Galileo rattled the cage, it was not until the Enlightenment that we had proven that the Sun was the centre of the Solar System. It was not until 1992 that the Catholic Church accepted the heliocentric conclusions of Galileo which were based on observations, not ideology.

The average person today "knows" that the Sun is at the centre of the Solar System but cannot prove it by astronomy and mathematics. Some people know that human emissions of CO_2 drive climate change but also can't prove it from science. Readers can be comforted by the fact that climate "scientists" also can't demonstrate that human emissions of CO_2 drive climate change.

The astronomer Ptolemy wrote geographic tomes with maps showing topography, used a grid pattern for the globe and constructed astronomical tables. His studies led to surveying. Ptolemy's geocentric view of the Solar System predominated from the second century AD until the mid-19th century. One of the reasons it took so long to reject the geocentric view was that the science was settled and was underpinned by religious authority. History has shown us that settled science only impedes the advance of science. European scientific thinking in the 16th century was dominated by re-discovery of the works of the ancient Greeks and not new scientific research. This was not an advance of science.

The Church challenges science

Heliocentrism was revived by Nicolaus Copernicus (1473 to 1543 AD). He created a model of the Solar System whereby planets rotated around the Sun.[96] Copernicus was the father of the scientific revolution that continued for centuries. After Copernicus, Tycho Brahe (1546 to 1601 AD), who was not a heliocentrist, and Johannes Kepler (1571 to 1630 AD)[97] led astronomic studies in Europe for the next century although there were only about 15 astronomers who accepted the heliocentric theory of the Solar System. Pope Gregory XIII used heliocentric concepts in 1582 AD to change from the Julian calendar to the current Gregorian calendar.

Galileo Galilei (1564 to 1642 AD) improved the telescope, a Dutch invention, and from astronomical observations was able to show that the Sun was the centre of the Solar System. He was also interested in tides and comets. This heliocentric view was challenged by other astronomers who argued that the Earth was the centre of the Solar System or supported the Tychonic system where it was argued that the Earth was the centre of the Universe, the Sun and the Moon rotated around the Earth and the other five known planets rotated around the Sun.

In 1610 AD, Galileo published his telescope observations of the phases of Venus and the moons of Jupiter.[98] The Rome Inquisition in 1615 AD concluded that Galileo's idea of heliocentrism was a

---

96: *De revolutionibus orbium coelestium* (1543 AD)
97: Johannes Kepler *Epitome astronomiae Copernicanae* (1617-1621 AD)
98: *Sidereus Nuncius* (1610)

possibility but not an established fact. Telescope observations led to the conclusion that the planets rotated around the Sun. This was published[99] and was misinterpreted as an attack on Pope Urban VIII. Previous support from the Pope and the Jesuits for Galileo evaporated. After a trial by the Inquisition in 1633 AD for his support of heliocentrism, Galileo was found to be suspect of heresy, forced to recant and spent his last years under house arrest.

It was argued that Galileo's theory was contrary to the Holy Scripture, Galileo was denounced from the pulpit and denounced for Copernicanism as well as a few other heresies supposedly spread by his pupils. At the Inquisition, Galileo's inquisitors stated that the idea that the Sun is stationary is "foolish and absurd in philosophy, and formally heretical since it explicitly contradicts in many places the sense of the Holy Scripture … ".

Banned books

The Inquisition judgment document was only widely available in 2014 AD. Copernicus' writings were banned and, in 1633 AD, Galileo's best selling *Dialogue Concerning the Two Chief World Systems* (1632 AD) was also banned. In 1758 AD, the Catholic Church dropped the prohibition of heliocentrist books from the List of Prohibited Books[100] that included publications deemed heretical, anti-clerical or lascivious. It was not until the next edition in 1835 AD that Copernicus' *Revolutionibus* and Galileo's *Dialogue* were omitted. The 20th and final edition of the List (or Index) was printed in 1948 and it was formally abolished by Pope Paul VI in 1966 AD.

In 1992, Pope John Paul II finally vindicated Galileo:[101]

> *Thanks to his intuition as a brilliant physicist and by relying on different arguments, Galileo, who practically invented the experimental*

99: *Dialogue Concerning the Two Chief World Systems* (1632 AD)
100: *Index Librorum Prohibitorum*
101: *L'Osservatore Romano* N. 44 (1264), 4 November 1992

method, understood why only the sun could function as the centre of the world, as it was then known, that is to say, as planetary system. The error of the theologians of the time, when they maintained the centrality of the Earth, was to think that our understanding of the physical world's structure was, in some way, imposed by the literal sense of Sacred Scripture …

The geocentric view of the Solar System was settled science for 1,850 years and those that promoted a heliocentric view were persecuted and made outcasts. Powerful institutions promoted a view that was contrary to evidence. The words "scientific consensus" or "settled science" must be viewed in this context.

Philosophiæ Naturalis Principia Mathematica

Isaac Newton (1642 to 1726 AD) made contributions to optics, classical mechanics, gravitation and the laws of motion.[102] He built reflecting telescopes, developed a theory on the spectrum of light, studied cooling and the speed of sound, tried to understand how fluids moved and developed new methods of mathematics. Newtonian physics was settled science until the 1840s when it was observed that the orbit of Mercury was non-Newtonian. This paradox was explained by postulating a planet between Mercury and the Sun. There was no such planet. Newton had noticed popular delusions and once commented: "*I can calculate the motion of heavenly bodies but not the madness of people*". This is still true today.

Newton assumed space to be everywhere and all the same and it was not until 1915 that Albert Einstein, with his theory of general relativity, showed that space is distorted by large objects such as the Sun. One of the early proofs of relativity was to calculate Mercury's orbit. It was exactly where Einstein's theory predicted. The "settled science" of Newton was shown to be incorrect after 230 years. There has never been "settled science". As soon as there is a

---

102: *Philosophiæ Naturalis Principia Mathematica* (1687 AD)

claim of "settled science", then we should instantaneously conclude that the wool is being pulled over our eyes.

In my university life, a number of us often joked that a great scientist is defined as a person who has held back advances in his field for as long as possible. This is done by reviewing and rejecting competing papers, acting as editor and not even sending a competing paper to reviewers, sitting on promotion committees, sitting on grant committees, reviewing research grant applications and using every mechanism to stop a contrary idea from an up-and-coming competitor seeing the light of day. I can think of many in my field just like this (and you know who you are). The same applies for all other fields of science and has been made an art form in climate "science".

Evidence-based thinking

We know what we know because of evidence. Science is married to evidence. There are always surprises and it is very difficult to make meaningful predictions. Evidence is collected by observation, measurement and experiment. Computer models do not constitute evidence. They are a method of trying to navigate through and understand a large body of data. Computer models are not the only way of trying to understand data. Although computer models have been used to try to make predictions, nature is very fickle and computer models suffer from 'unknown unknowns.' There is a view in society that a belief is evidence and a strongly held belief must be correct. This is emotion, not evidence. Most public discussions about climate change are underpinned by emotion and normally the intensity of emotion is inversely proportional to the amount and veracity of evidence.

In science, the debate is about whether the process of evidence collection was valid, what the errors were, what was the order of accuracy, what assumptions were made and whether such assumptions were valid. When I look at a scientific work, I like to know who collected the data, when it was collected and where it was collected. Much scientific data is collected by assistants and students and then given to the scientist to analyse. What happens if the primary data is incorrect? Much climate "science" research comprises a mathematical and computer model analysis of other people's data. In most cases, this data cannot be independently validated.

When it is judged that there is a suitable body of evidence, this body of evidence is interpreted and explained. Sometimes, data is judged unreliable and has to be recollected. There are no hard and fast rules for such judgments. Furthermore, some material just cannot be recollected (e.g. lunar, Martian, asteroid and space samples) and scientists have to be content with what data is available. The interpretation and explanation of a body of data forms a scientific theory. In the law, some evidence may be inadmissible which is why some legal decisions seem bizarre. Not so in science.

Some scientific ideas such as cold fusion are very quickly abandoned when the phenomenon can't be reproduced and validated. The real peer review process actually occurs after publication of a paper. Very often green left environmental activists have too much to say about the peer review system and this only demonstrates that these scribes have never had work peer reviewed, have never been a reviewer and have never been an editor. The models have another fundamental error. They assume that in 100 years time the world will not be much different from the baseline (1990). The world of 2015 is already very different from the world of 1990 and change will continue to occur on our dynamic planet. A reasonable person would argue that after waiting for over 20 years for the planet to warm, the most logical conclusion could be made. The logical conclusion is that the prediction of human-induced global warming is demonstrably wrong.

Scientific evidence must be reproducible. Science transcends cultures, religion, gender and race. The scientific theory will change with

new validated evidence hence it is impossible in scientific discipline
for the science to be settled. Science is anarchistic, has no consensus,
bows to no authority and it does not matter what a scientific society, government or culture might decide, only reproducible validated
evidence is important. There have been exceptions such as Lysenko
and climate "science". Like any other area, science suffers from fads,
fashions, fools and frauds, has short-term leaders and can be cult-
ish.

In my scientific career, it took me a little time before I refused to
follow fashions and fads, married my opinion to reproducible validated evidence and did not become a member of any putsch pushing
a particular scientific theory. At any scientific conference, one can
see the various putsches prancing around narcissistically competing
for attention, fame and research grant fortune while espousing their
competing theories and demonising opponents. This is normal human behaviour. Scientists do not hold the high moral ground and
have the same weaknesses as their fellow man.

Scientific ideas can be tested. There may be one hundred sets of
evidence in support of a scientific idea. It only takes one piece of validated evidence to the contrary and that idea must be rejected. That
is Karl Popper's concept of falsification or refutation.[103] Adolf Hitler
sought to discredit the work of Albert Einstein by commissioning
a paper[104] called *"100 Authors Against Einstein."* Einstein responded:
"… it doesn't take 100 scientists to prove me wrong, just one fact will do …".
Criticism of Einstein was by 28 authors, excerpts of publications of
19 others and a list of others who at some time were not convinced
of relativity. This is rather like having a list of scientists and professional societies who support human-induced global warming. The

---

103: Popper, K. 2005: *The logic of scientific discovery*. Taylor and Francis
104: Israel, K. *et al.* 1931: 100 Autoren gegen Einstein. *Naturwissenschaften* 19:
254-256

Royal Society comes to mind, which is contrary to their motto[105] *Nullius in verba*. The human-induced global warming farce has evolved in a similar fashion. Creationism has the same characteristics with long lists of allegedly eminent titled folk with impressive looking post-nominal sheep skins.

Manipulation of data

To change primary or raw data in science is a cardinal sin. It is almost a universal procedure in climate "science". That's how conclusions can be reached that are unrelated to other areas of science. The papal advisors should be aware of the manipulation of data and climate "science" should have been roundly criticised for leading the world astray. They were not and the Encyclical makes no mention of scientific fraud.

The measured United States Historical Climate Network (USHCN)[106] daily temperature data shows a decline in air temperature in the USA since the 1930s. Before this data is released to the public, it undergoes a number of "adjustments" which just happen to change the trend from cooling to warming.[107] The difference between raw data and "adjusted" data sets shows that the past is cooled and the present is warmed. No matter what temperature is really doing, this shows exaggerated warming. Such skullduggery is not restricted to the US. The UK's Met Office's Hadley Centre is responsible for compiling the HadCRUT global temperature datasets of land and sea surface temperatures in collaboration with the Climatic Research Unit at the University of East Anglia.

There is a land temperature data set, CRUTEM, derived from air temperatures measured at weather stations near to the land surface

---

105: Take no one's word for it
106: http://cdiac.ornl.gov/epubs/ndp/ushcn/ushcn.html
107: https://stevegoddard.wordpress.com/data-tampering-at-ushcngiss/

across all continents of the Earth. The CRUTEM4 data set shows a 2.2°C temperature rise per century. When the data set was changed to CRUTEM4.3, the temperature rise had changed to 2.8°C per century. There was no explanation.

The HadCRUT 4.3 data set was changed to the HadCRUT 4.4 data set.[108,109] Without collecting new data there have been "adjustments" to the historical record wherein the past is cooled and the present is warmed,[110] the end result of which is to remove the >18 year pause in warming. It is a cardinal sin in science to adjust raw data. This is fraud. This data was paid for by the taxpayer and should remain as raw data for perpetuity. Tampering with Goddard Institute for Space Studies (GISS) air temperature data over the last five years has suddenly retrospectively shown that instead of cooling, trends are now reversed.[111,112] Requests for explanations have gone unanswered.

Recent temperatures have been "adjusted" upwards by 0.2°C. Hence, since the warm 1930s and 1940s, the adjusted temperatures show an increase of 0.6°C. The adjustments account for at least one third of the post 1930 warming. With every adjustment, the warmest year on record shuffles up or down the ranking of the warmest year. This is fraud.

In 2014-2015, The National Oceanic and Atmospheric Administration (NOAA) has rewritten the climate history of Maine twice. When asked for an explanation, they stated:[113]

108: http://www.cru.uea.ac.uk/cru/data/temperature/HadCRUT3-gl.dat

109: http://www.metoffice.gov.uk/hadobs/hadcrut4/data/versions/previous_versions.html

110: http://notalotofpeopleknowthat.wordpress.com/tag/temperature-adjustments/

111: http://data.giss.nasa.gov/gistemp/

112: http://wattsupwiththat.com/2015/07/24/impact-of-pause-buster-adjustment-on-giss-monthly-data/

113: http://manhattencontrarian.com/blog/2015/7/21/the-greatest-scientific-fraud-of-all-time-part-v/

... improvements in the dataset, and brings our value much more in line with what was observed at the time. The new method used stations in neighboring Canada to inform estimates for data-sparse areas within Maine (a great improvement).

What this means is obvious. Primary data has been tampered with to produce pre-ordained results. The world temperature data record is already a hybrid of raw data, tampered data and estimates. There is now very little reliable temperature data for future generations.

For years, GISS have been gradually "adjusting" historical records in order to increase the warming trend. Raw data is "adjusted" to show that the current year is the hottest one on record and there is an uncritical media fanfare. Raw data is "adjusted" to show that the >18 year pause in warming is not really a pause because the "adjusted" raw data now shows warming. Again, a big fanfare from the uncritical compliant media echo chamber. No wonder the GISS data is now diverging from the more accurate satellite data.[114] If this was in the commercial world, there would be a prison cell waiting. It is very much a sign of how green left environmental activists have captured taxpayer-funded academia and feel free to conduct fraud to promote an ideology.

Data "adjustments" have occurred at Reykjavik (Iceland), De Bilt and Uccle (Netherlands),[115] Trier and Hannover (Germany), Alice Springs, Darwin, Bourke, Brisbane, Wagga, Deniliquin, Kerang and Rutherglen (Australia).[116] Dr Jennifer Marohasy has shown that the Australian Bureau of Meteorology just makes up an "adjustment". They claim that moving a measuring station between paddocks near a small town in north-eastern Victoria (Rutherglen, Australia) re-

114: http://www.woodfortrees.org/plot/gistemp/from:1998/to:2015/plot/gistemp/from:1998/t):2015/trend/plot/rss/from:1998/to:2015/plot/rss/from:1998/to:2015/trend
115: https://notalotofpeopleknowthat.wordpress.com/2015/04/09/coolin-the-past-in-holland/
116: http://wattsupwiththat.com/2009/12/08/thesmoking-gun-at-darwin-zero/

quired a change from a cooling trend of 0.35°C per century to a warming trend of 1.7°C per century.[117] The trend was "adjusted" in order for it to be in accord with "adjusted" trends at neighbouring weather stations.

There may be a yet to be a reason to "adjust" raw data but, despite requests, no explanations have been given. It appears that long-term temperature records have been "adjusted" to the point of unreliability. Taxpayers fund the Bureau of Meteorology to keep accurate long-term records of raw data yet it appears that "adjustments" are made for ideological reasons. If it is in order for the Bureau of Meteorology to "adjust" raw data, then it is perfectly logical for governments to adjust the budget of the organisation. It is no wonder that the BBC have sacked the Met Office as the source of their weather information and will have a New Zealand company do the job.[118]

All "adjustments" seem to go the same direction and exaggerate warming. Where is the "adjustment" that shows cooling? Surely with so many "adjustments" there would be just one lonely "adjustment" that was changed from warming to cooling. After decades collecting and interpreting raw data, this looks to me like systematic large-scale fraud by scientists who benefit from the catastrophist global warming gravy train. The land-based air temperature measurement data is now not reliable and it would be reasonable to conclude that some data used for models is corrupted and altered beyond usefulness. The enigma of global warming has finally been solved. It is mainly due to fiddling the raw data. You heard it here first.

In 2015 there is a flurry to change the raw land-based temperature data (HadCRUT4, GISS, NOAA) to show exaggerated warming in the build up for the Paris climate conference (whose true purpose is to establish an unelected and all-powerful global "governing body"). The

---

117: http://jennifermarohasy.com/2015/08/bureau-just-makes-stuff-up-deniliquin-remodelled-then-rutherglen-homogenized/
118: *Sunday Telegraph*, 30 August 2015

satellite data (University of Alabama at Huntsville [UAH], Remote Sensing Systems [RSS]) has not been amended yet. It won't be long.

The Pope is concerned about warming. Did one of his advisors ask simple questions about air temperature measurement and whether the primary data was "adjusted" by various meteorological bodies or the keepers of the data such as HadCRUT or GISS? There is an extensive literature about the greatest scientific fraud of all time: the "adjustment" of primary data by US and UK government employees from NOAA and the Hadley Centre.[119] NOAA puts out press releases more or less on a monthly basis about higher and higher temperatures and new "news" with even scarier headlines yet fail to mention that other data sets do not show the same records or warming and just never seem to mention the serious, unrefuted and proven allegations of data tampering and fabrication.

A little bit of thought by the Pope's advisors would have made them uneasy about those who keep the records and run the models that predict the scary scenarios. They are the very same people responsible for the databases that validate the predictions. Is this not a conflict of interest? Where is the transparency and dispassionate collection of data? Had the Pope's advisors done enough reading to be sceptical about temperature, measurements, adjustment of data and media stories about predicted doom and gloom? If temperature data is "adjusted", how can we rely on other data and conclusions?

This is a moral issue ignored by the Pope.

Going off the rails

At times, science goes off the rails. We are in one of those times. In the past we saw it with eugenics and Soviet agricultural policy allegedly underpinned by science. History shows us that it takes decades to centuries to get back on track.

119: http://www.informath.org/WCWF07a.pdf

Eugenics

There was once a scientific consensus about eugenics. This was peer reviewed by only the best people. The plan was to identify those that were feeble-minded (and that variously included Jews, blacks and foreigners) and stop them from breeding by isolation in institutions or by sterilisation. At times, eugenics research was funded by the Carnegie Foundation and the Rockefeller Foundation. With some eminent citizens and respectable foundations supporting eugenics, how could it not be good science?

We now know it was a racist, anti-immigration social program masquerading as science. So too with human-induced global warming. When so many allegedly important scientists and professional societies supported the theory, how could it be wrong?

Lysenko lies

History has an unfortunate habit of repeating itself. Government ideology and an uncritical media have led to the promotion of false scientific concepts. One example was Trofim Denisovich Lysenko (1898-1976 AD), a self-promoting Russian peasant who invented a process called vernalisation. This was the time of Stalin in the USSR. Seeds were moistened and chilled to enhance later growth of crops without fertiliser and minerals. It was claimed that the seeds passed on their characteristics to the next generation. Stalin wanted to increase agricultural production, vernalisation never had scientific scrutiny and didn't involve the cost of fertiliser.

Lysenko became the darling of the Soviet media and was portrayed as a genius and any opposition to his theories was destroyed. Does Michael Mann's hockey stick come to mind? Lysenko's theories dominated Soviet biology for decades and the effects still can be seen. Vernalisation did not work. Millions died in famines, hundreds of dissenting scientists were sent to the gulags or the firing squads and genetics was called the *bourgeois pseudoscience*. The evidence-based

Mendelian genetics was rejected in favour of an ideology (Michurin's hybridisation). The plummeting of grain production resulting in the deaths of more than 14 million people between 1929 and 1933 was a direct result of Lysenko. More people were killed by Stalin's Lysenkoism and collectivisation of agriculture than were killed in World War I or Hitler's genocide.[120]

The soft cuddly side of warmists

The same savage criticism occurs against scientists who challenge the theory that human emissions of CO_2 drive global warming on the basis of evidence. Some extremist environmentalists are suggesting that such scientists should be removed from employment, imprisoned and even assassinated. For example:[121]

> *On September 20th the British newspaper, The Guardian, published an article on a letter from the Royal Society ... calling for the silencing of groups, organizations, and individuals who do not conform to their views on climate change and policy.*

and[122]

> *The British Greens have called for a purge of officialdom to get rid of anyone who doesn't accept 'scientific consensus on climate change'.*

Other scribes have had their two bob's worth on the same theme:[123,124,125]

> *I wonder what sentences judges might hand down at future international criminal tribunals on those who will be partially but directly*

120: Robert Conquest, 1987: *Harvest of Sorrow: Soviet collectivization and the Terror-famine*. Oxford University Press
121: Kueter, Jeffrey, 29 September 2006. President of George C. Marshall Institute, letter to Congress
122: O'Neill, Brendan, *The Weekend Australian*, 2-3 May 2015
123: Lynas, Mark, *Dagelijkse Standaard* 19 May 2006
124: Monbiot, George, *The Guardian*, 23 August 2008
125: Lovelock, James, *The Guardian* , 29 May 2010

responsible for millions of deaths from starvation, famine, and disease in the decades ahead. I put [their climate change denial] in a similar moral category to Holocaust denial – except that this time the holocaust has yet to come and we still have time to avoid it. Those who try to ensure we don't will one day have to answer for their crimes. (Lynas, 2006)

and

Stopping runaway climate change must take precedence over every other aim. Everyone in this movement knows that there is little time: the window of opportunity in which we can prevent two degrees of warming is closing fast. We have to use all the resources we can lay out hands on, and these must include both governments and corporations. (Monbiot, 2008)

and

I feel that climate change may be an issue as severe as war. It may be necessary to put democracy on hold for a while.(Lovelock, 2010)

Professor Richard Parncutt, a professor of systematic musicology, at Karl-Franzens-Universität Graz (Austria) has clearly critically evaluated the science of human-induced global warming and gave us his expert opinion:[126]

As a result of that process, some global warming deniers will never admit their mistake and as a result they will be executed. Perhaps it would be the only way to stop the rest of them. The death penalty would have to be justified in terms of the enormous numbers of saved future lives.

Kevin Trenberth, one of the key fraudulent figures in climategate, had 19 climate "scientists" join him in a 1 September 2015 letter to

---

126: Parncutt, Richard, 25 October 2012, *Death penalty for global warming deniers? An objective argument ... a conservative conclusion.* Internet text on website of Karl-Franzens-Universität Graz (Austria) until removed by order of university officials

President Obama demanding that those that disagree with them be gaoled under the *Racketeer Influenced and Corrupt Organisations Act.* It appears that Trenberth *et al.* are tired of debating global warming and want dissidents incarcerated.

I wonder if these murderous and fraudulent anti-democratic scribes would be prepared to take the chop if they are wrong and the planet cools. They are the deniers who try to tell us that there was no Medieval Warming, that there has been atmospheric temperature rise during the last 18 years and that there is a correlation between atmospheric CO_2 and temperature. Green left environmentalist activism is the new Lysenkoism and such movements are a honey pot for all sorts of despots and nutters. Many of them also proudly call themselves socialists.

It is these modern Lysenkoists that the Pope has embraced.

Abandoning treasured ideas

The surge in populist global warming anti-science reminds me of the breatharians who claim that they only need air and energy from sunlight, don't need nutrition from food and fluids and claim that their nutrition comes from cosmic micro-food. It appears that through meditation, one can obtain all the nutrients for life from air and sunlight. Some of them have been deluded enough to be tested under laboratory conditions and then give up after a few days of dehydration, dizzy spells, confusion and emaciation. There is, of course, always an excuse as to why the phenomenon did not work in that specific laboratory test and the breatharian ideology is not abandoned. Normally the excuse is that such phenomena don't work in the presence of sceptics or under laboratory test conditions.

There have even been breatharians who have starved themselves to death,[127,128] a great example of Darwinism for which, shamefully, a Darwin Award has never been given. In 1999, a lady took to the wilds of the Scottish Highlands with a tent and her breatharianism. She was found dead from hypothermia and dehydration aggravated by a lack of food. Tragically, she only received a Darwin Award nomination.[129] She certainly improved the human gene pool by removing herself from it.

So too with the hot air climate catastrophists who cling onto an irrational ideology despite a lack of scientific data to show that human emissions of CO_2 drive dangerous global warming. Although climate models are demonstrably wrong and there has been no measurable increase in global atmospheric temperature for over 18 years despite an increase in atmospheric and human emissions of CO_2, they claim that this is not a valid test and that in 50 or 100 years they will be shown to be correct.

This treasured idea is easily tested by comparing it with the reproducible validated evidence from past climate changes. This is the coherence criterion of science.[130] Any new idea needs to be in accord with validated evidence from other areas of science. Human-induced global warming is not in accord with geology. Nor is it in accord with history. There are hundreds of examples that show that past events of global warming were not driven by CO_2, that the planet has been far colder and warmer in past times and that the rates of temperature change have been far faster than any changes measured today.

127: "Swiss woman dies after attempting to live on sunlight; Woman gave up food and water on a spiritual journey" *Associated Press*, 25 April 2012
128: "Three deaths due to 'living off air' cult." *Sunday Times*, 26 September 1999
129: http://darwinawards.com/darwin/darwin1999-58.html
130: Evers, C. W. and Lakomski, G. 1996: *Exploring educational administration: Coherentist applications and critical debates.* Pergamon

This data, which invalidates the theory that human emissions of CO_2 drive climate change, is not even acknowledged by the green left environmental activists. This shows that the idea of human-induced global warming is not science and is a political ideology. The past is a story of constant climate change and our present climate is the result of the past and continues into the future. The idea that human emissions of CO_2 that lead to catastrophic global warming is therefore invalid and to continue to promote such an idea is ignorance, fraud or perhaps both. No wonder those that call themselves climate "scientists" don't want to debate geologists. The climate "scientists" solution to this spot of bother is to ignore geology, astronomy and history and yet claim that they are involved in science. All this shows is they are involved in politics and not science.

Uncertainty, certainty and consensus

As a scientist I can write definitely that I am not sure whether there is certainty or uncertainty in science.

Uncertainty

All scientific evidence and the ideas created from this evidence have a degree of uncertainty. To make strong statements or predictions without declaring the uncertainties is not science. Geologists learn this very quickly. Testing of ideas based on the latest complex geophysics, geochemistry and geology hidden below with a drill hole is a very humbling experience. We often jokingly call a diamond drilling rig a rotary lie detector. Models based on geophysical measurements are most commonly shown to be wrong by drilling. The new subsurface evidence from drilling and measurements on the diamond drill core are used to refine the geophysical model. The model may still be wrong but it is better than the first attempt and can be tested.

For scientists to declare certainty, to make predictions and to ignore large bodies of contrary science shows that personality weaknesses overwhelm the scientific method.

Scientific certainty and 97% consensus

The whole issue of human-induced global warming has become a gravy train. Even sociologists have swooped in to get funds to study why people may be sceptical about human-induced climate change. For example:[131]

> *Climate scepticism persists despite overwhelming scientific evidence that anthropogenic climate change is occurring (IPPC, 2013)[sic]. The reasons for this are varied and complex. Understanding why climate scepticism endures or is even on the rise, and why levels of scepticism vary across countries, requires accounts that recognise that biased information assimilation is mediated by cross-cultural and intra-national differences in values and worldviews. Fruitful explanations of scepticism must also account for the way in which partisans are influenced by their political leaders. Integrating such accounts may provide a way to both understand and address the social problem of climate scepticism.*

This is post-modernist gobbledygook. Climate always changes hence there is no such thing as climate scepticism. There is not overwhelming evidence that anthropogenic climate change is occurring, the only reference used to support this statement being by an activist group with self-interest. There is a huge literature in the integrated interdisciplinary world of science (as I outlined in *Heaven and Earth. Global warming: The missing science*)[132] that has well-founded but different conclusions. One doubts whether sociologists Tranter and Booth read the scientific literature let alone understand it.

They certainly did not read the IPCC report, although they use

---

131: Tranter, B. and Booth, K. 2015: Scepticism in a changing climate: A cross-national study. *Global Envir. Change* 33:154-164
132: Plimer, Ian 2009: *Heaven and Earth. Global warming: The missing science.* Connor Court

an unknown IPPC 2013 report [sic] as a reference so presumably they have read this report because they claim it provides "*overwhelming scientific evidence*". In places, the IPCC report shows great reserve and shows that the magnitude of human influence on global climate is unknown.[133,134,135,136] The sociologists noted that when identifying "climate sceptics", their study did not mention human-induced climate change. Their world-shattering leading edge research was a survey. The survey questions were trite, nonsensical and made unsubstantiated assumptions. The methodology of the survey from which conclusions are based is flawed. Natural climate changes and human-induced change (e.g. due to land clearing, urban heat island effect) take place yet there is still no data to show that human-induced global climate change from human emissions of CO_2 is significant or dangerous. A flippant concluding reference to an IPPC 2013 [sic] report shows the standard of scholarship, peer review and editorial rigour.

The reason why scepticism is on the rise is that the average taxpayer has realised the science is exaggerated, embellished, wrong or fraudulent, that time and time again models and predictions have been wrong and the consequences of failed politicised science have

---

133: IPCC 2013, SPM, p. 3, Section B.1, bullet point 3, and in Synthesis Report p. SYR-6 ("... the rate of warming over the past 15 years...is the rate calculated since 1951 ...")

134: IPCC 2013, WGI, Ch 9, box 9.2, p. 769 and in Synthesis Report SYR-8 ("... an analysis of the full suite of CPIP5 historical simulations...reveals that 111 out of 114 realisations show a GMST trend over 1998-2012 that is higher than the entire HadCRUT4 trend ensemble ...")

135: IPCC 2013, SPM, D.1, p.13, bullet point 2 and Synthesis Report SYR-8 ("There may also be a contribution from forcing inadequacies and, in some models, an overestimation of the response to increasing greenhouse gas and other anthropomorphic forcing [dominated by the effect of aerosols]")

136: IPCC 2013, WGI, ch 9, box 9.2, p. 769 ("This difference between simulated [i.e. model output] and observed trends could be caused by some combination of (a) internal climate variability, (b) missing or incorrect radiative forcing and (c) model response error.")

started to become very expensive for the average person. They have become sick of being told that the world's going to end. The green left environmental alarmists have been crying wolf for a quarter of a century, the climate has not changed noticeably, people are struggling even more to pay their energy bills and the community now no longer listens to those who cry wolf.

People have not been influenced by their political leaders. Political leaders have responded to the community scepticism and acted accordingly. In the UK, Australia and elsewhere some people have changed political allegiances because a particular party has uncritically embraced a policy on climate change. Maybe Tranter and Booth have a naïve understanding of politics and don't understand that many people cast their vote on a diversity of policies rather than down party lines.

Tranter and Booth label climate scepticism as a social problem hence they can rationalise opening the floodgates for further research funds. They forget to add their assumption in the paper: That climate change occurs and has been taking place for billions of years. Did the authors ever consider that scepticism underpins intellectual thought, that there is far more access to information today than ever before and the average person can now critically analyse claims made by green left environmental activists who claim that they are scientists? One does not have to be a titled person in a university to possess knowledge and the skills of critical analysis.

Independent thinkers such as sceptics do not accept being told what to think, need evidence to underpin doomsday predictions and do not accept pronouncements made upon high by authorities. They actually think for themselves. The authors comment that sceptics tend to be older. Maybe older people received a more rigorous education in which facts and thinking processes were paramount rather than feelings, beliefs and environmental ideology.

After nearly 45 years on the staff of various universities and a

Chair for nigh on 30 years, all I can state is that many current university staff have been trained in a system that has been dumbed
down. It shows. Some of them, although now middle-aged, tena-
ciously hang on to their undergraduate young and foolish Marxist
view of the world to make their arrested development complete. A
more fruitful line of sociology research would be to investigate why
many people are so gullible and accept the unsubstantiated claims
and models of the IPCC when there is such a large body of contrary
evidence.

Tranter and Booth are clearly gobsmacked as to why the average
taxpayer is not fooled. They need to get a life. They clearly treat with
great disdain anyone who is not a high and mighty academic sociolo-
gist and who follows the current fad. A word of advice to Tranter
and Booth: The public is not stupid. And what sort of journal pub-
lishes this tosh. The quote above shows that this article is certainly
not scholarship, the facts were not checked, the refereeing was poor
and the editing does not seem to exist.

Sociology journals have form. How can we forget that the physicist Alan Sokal submitted a sham article generously salted with
nonsense, pseudo-babble and obvious howlers to the cultural studies journal *Social Text?* It was published in 1996.[137] The article was
submitted to show the decline in standards, especially in the soft
post-modernist bleeding heart hand-wringing disciplines. Sokal exposed his article himself; it was not discovered by readers, reviewers
or editors.[138] The journal editors tried to justify publishing the article
because it was "the earnest attempt of a professional scientist to seek
some sort of affirmation from postmodern philosophy for developments in his field."

137: Sokal, A. D. 1996: Transgressing the boundaries – Toward a transformative
hermeneutics of quantum gravity. *Social Text* 46/47: 217-252
138: Sokol, A. D. 1996: A physicist experiments with cultural studies. *Lingua
Franca* May/June 1996: 62-64

It was no such thing. It was a published paper showing the lack of scholarship in a major journal of sociology. One of the editors even claimed that "Sokal's parody was nothing of the sort and that his admission represented a change in heart or a folding of his intellectual resolve."

It was no such thing, despite the editors of *Social Text* making valiant weak explanations.[139] It was not a parody, it was an exercise to show the lack of scholarship in sociology. Sokal was not admitting to being dishonest or changing his view. He was a rigorous physicist who showed that some sociology journals accept whatever nonsense is submitted to them as scholarship couched in post-modernists deconstructionism language. The title of the paper was absolute nonsense and any educated person would have seen that the text was nonsense.[140] If I was a referee of the Tranter and Booth paper, I would have suggested to the journal editors:

> ... and with a bit more hard work the paper would be of the standard of Sokal (1996).

Some of us are unreconstructed sceptics. Some of us have a deep-seated scepticism for anything produced by activists, career academics, governments, political parties, fundamentalist religious

---

139: http://linguafranca.mirror.theinfo.org/9607/mst.html

140: A quote by Sokal: "The Einsteinian constant is not a constant, is not a center. It is the very concept of variability – it is, finally, the concept of the game. In other words, it is not the concept of something – of a center starting from which an observer could master the field – but the very concept of the game." I have read this many times with and without James Squire amber ale on board, it still means nothing to me. Another quote from the Sokal paper should have given the game away: "the pi of Euclid and the G of Newton, formerly thought to be constant and universal, are now perceived in their ineluctable historicity." I have been an editor of a major scientific journal for years and have seen some dreadful papers submitted. Why didn't the editor at least insist that such great revelations were intelligible? An editorial tip: If something has to be read more than once to be understood, then it is poorly written or nonsense.

organisations, big business, self-interest groups, various environmental groups and anyone who rejects data, logic and rationality for emotional arguments.

There is no substitute for common sense.

Scientific disputes

Every active scientist is constantly involved in scientific disputes because science is never settled on anything, new data and ideas are continually being aired and any new idea is generally critical of previous work. That is the nature of science. It evolves over time. Many of us have written papers with our critics and rarely do scientific disputes become personal. Some of my closest personal friends have been savagely critical of my work in their role as a referee in the peer review process. This is intellectual honesty and, although irritating at the time, does not change friendships.

Some of us have been editors of major scientific journals, sat on editorial boards, refereed scientific papers, sat as experts on research council panels, examined doctoral theses from all over the world, been called as expert witnesses in Court and are scientific authors. Disputes are normal. It is not for the media or one side of a polarised debate to decide whatever may be deemed the truth.

In my capacity as Managing Editor of *Mineralium Deposita*, I was presented with a paper by a prominent creationist's PhD work on the uranium ore deposits in the Northern Territory. The author used scientific terms for rocks billions of years old yet in the creationist literature, the same author argued that the planet was only a few thousand years old. Does one reject the paper because of the author's religious views which he claims are underpinned by science or does one send the paper to referees?

After reading the paper, I sent it to referees because, in my editorial and scientific judgment, the paper contained science that was of international interest. After very minor editorial changes, the paper was published. At the same time, I was engaged in battles with that very same creationist in the creationist literature and in the media. As a senior scientist, it's very easy to kill off the work of rivals, competitors and those that we dislike. In climate "science", it is universal.

Peer review

Peer review depends on the integrity of the editor and anonymous referees. If the editor has bias, it's easy to send a paper to peer reviewers who will give the desired expert opinions. A scientific paper or grant proposal can easily be buried by sending the work to a referee in another camp. As an editor, member of a research grant committee or member of an appointment committee, it is useful to know who are the players in competitive tribalism because an opposing clique often buries good work. Peer-review is only an editorial aid. It involves rejecting or accepting a paper and making suggestions and changes to the submitted paper. The number of reviewers that can be called on for a submitted paper is generally two or three people.

Peer-review is conducted at the discretion of editors and the peer-review process is very susceptible to the influence and bias of small groups promoting and protecting their own interests. Most reviewers are anonymous. This only serves to strengthen the influences, biases and cowardice via anonymity. Pre-publication review is important but the most important process occurs after review. This is when the scientific community at large can refute, check, confirm or expand on the ideas published.

The climate clan is peer reviewing and publishing their own grant-earning dogma and some of the scientific community is showing how weak this science is post-publication. Peer-review, far from being the gold standard, is used by the climate industry to approve or censor science according to its own agenda. In the climate industry, it is not peer review, it is peer bullying and peer pressure to conform to the orthodoxy. The work of Newton, Darwin and some of that of Einstein was not peer-reviewed and it is wrong to impute that only good science is peer-reviewed.

After a lifetime of science, I have seen lots of shoddy published science and failings in the peer-review system. For example, Jan-Hendrik Schön managed to publish a paper every eight days in major scientific journals between 2000 to 2002 on nanotechnology and single molecule behaviour. Every paper was peer-reviewed, every paper was bogus and yet Schön's science was published by major journals such as *Nature, Science, Applied Physics Letters* and *Physical Review*. Schön won a number of prestigious prizes for his published work. This fraud was later detected by a student and not by the peer-reviewers. This fraud was published in journals such as *Nature* and *Science* that have taken a partisan position on human-induced global warming.

Scientific journals should take no position on any issue, they should be indifferent to politics, should publish the spectrum of competing hypotheses and leave it to the market (i.e. the scientific community) to distil, validate or replicate over time. Schön showed that if you have influential mainstream mates, fraud could easily be published in the best peer-reviewed scientific journals. This was corrected by the post-publication evaluation of peer-reviewed papers. Peer review is not the gold standard that assures the quality of science publishing, which is why a number of journals and institutions have a science citation index that quantifies the scientific impact of a paper. Again, this is not the gold standard because in a small field the impact can be different from a big field and the various tribes are normally unwilling to cite the work of competitors.

A lot of what is published is incorrect. These are not my words. In medicine, maybe half of what is published is untrue. Studies with small sample sizes, tiny effects, invalid analyses, flagrant conflicts of interest, an obsession with pursuing fashionable trends of dubious importance seem to be the norm. Poor scientific methodology gets results and produces compelling stories. This is until they are checked or an attempt at replication is undertaken. There is so much concern about published incorrect medical sciences that the Academy of Medical Sciences, Medical Research Council and the Biotechnology and Biological Sciences Research Council have now put their reputational weight behind such concerns. Statistical fairytales abound, journals have poor peer review and editing processes and processes of granting research monies all contribute to bad scientific practices. If this is the case for the medical sciences,[141] what about other fields. I doubt it is any different.

There have been some high profile errors in physics (e.g. cold fusion, particle physics) that have resulted in changing procedures such that intensive checking and rechecking of data takes place prior to publication. What about climate "science"? I suspect that with Climategate, failed climate models, failed predictions, exaggerations, omission of contrary data and unfounded statistical studies, this figure would be far higher because I suspect that the fields of physics and medicine are more rigorous than climate "science". Your guess is as good as mine but it's not a pretty sight. Maybe with climate "science" it's a case of never let the science get in the way of a hefty research grant.

However, in the climate industry there are no parallels. There appears to have been a profession-wide decision that there can be nothing published that might threaten their fame- and fortune-giving ideology that global warming is driven by human emissions of CO_2. This lack of scientific objectivity is such that one has to question whether any literature from the climate industry can even be treated seriously for another 30 years. The climate industry has openly and flagrantly violated every aspect of the core ethos of science. Climate industry "science" is advocacy in the service of one arm of politics,

---

141: Horton, R. Offline: What is medicine's 5 sigma? *The Lancet* 385: 1380

is not dispassionate and is no more valid than the tobacco industry publishing science about the harmlessness of smoking.

There are many ways of showing that at times in the past, planet Earth experienced a warmer climate than today. This destroys the catastrophist narrative. The spectacular finding in 1998 that rising temperature trends were much higher today than in the past by Michael Mann[142] underpinned political action on human-induced climate change. Mann showed that rather than climate oscillating between warm and cold periods, global warming was now out of control and looked like a hockey stick. Mann's redrawn temperature graph was based on tree rings. The evidence and the statistical methodology were re-evaluated.[143] The hockey stick was a myth and a fraudulent creation *ex nihilo*.

This controversy led to a US Senate Committee setting up an inquiry under Professor Edward Wegman. The inquiry showed that Mann's work could not be supported. In the Wegman Report,[144] a scathing review of the fraud of the Michael Mann hockey stick, the peer review process in the climate industry was questioned:

> *One of the interesting questions associated with the "hockey stick controversy" are the relationships among authors and consequently how confident one can be in the peer review process. In particular, if there is a tight relationship among the authors and there are not a large number of individuals engaged in a particular topic area, then one may suspect that the peer review process does not fully vet papers before they are published.*

142: Mann, M. *et al.* 1998: Global-scale temperature patterns and climate forcing over the past six centuries. *Nature* 392: 779-787

143: McIntyre, S. and McKitrick, R. 2003: Corrections to the Mann et al. (1998) proxy data base and Northern Hemisphere average temperature series. *Energy Envir.* 14: 751-771

144: U.S Congress House Committee on Energy and Commerce; http://republicans.energycommerce.house.gov/108/home/07142006_Wegman_Report.pdf

and

Indeed a common practice among associate editors for scholarly journals is to look in the list of references for a submitted paper to see who else is writing in a given area and thus who might legitimately be called on to provide knowledgeable peer review. Of course if a given discipline area is small and the authors in the area are tightly coupled, then this process is likely to turn up very sympathetic referees. These referees may have co-authored other papers with a given author. They may believe they know that author's other writings well enough that errors can continue to propagate and indeed be reinforced.

The Wegman Report had only dealt with one small aspect of peer review. In the case of the "hockey stick", there were only 43 in the climate industry who could have been gatekeeper in the peer review process of Mann's fraud. With such a small club, it is not surprising that shoddy science can be published. Later work[145] has shown just how a small group of climate "scientists" managed to influence world thinking, especially amongst non-scientific green left environmental activists. This was until the "hockey stick" was proven to be fraudulent. The IPCC gave Mann's "hockey stick" star billing until it was shown to be fraudulent and then quietly got rid of it.

Over quite a period of time Ross McKitrick nibbled at Mann, his "hockey stick" and the IPCC's blind acceptance of Mann's fraud. McKitrick had great difficulty publishing papers in the mainstream climate "science" literature refuting claims by Mann regarding his "hockey stick" and the IPCC claims that real-world data confirmed models predicting man-made global warming. Ranks closed and the cosy catastrophic climate change club was protected. Ultimately, McKitrick was able to show that this was not a minor academic spat in the hallowed halls of academe but was symptomatic of the whole

---

145: Steyn, Mark 2015: *A disgrace to the profession. The world's scientists in their own words on Michael E. Mann, his hockey stick and their damage to science.* Vol. 1 http://www.steynstore.com/product113.html

climate industry closing ranks to protect their own, to inhibit free discussion, to hide or censor data and to stop data that leads to a different conclusion being published.[146,147,148,149]

Developments in climate "science" that challenge the findings of the IPCC continue. A 2011 paper by one of the world's most eminent meteorologists, Richard Lindzen, addressed concerns raised by critics of his 2009 paper. His research finds that a doubling of CO_2 will increase temperatures by only 0.7°C. This is significantly below the lowest estimates of the models used by the IPCC. The climate industry closed ranks and it took about two years for Lindzen to find a journal willing to publish the paper.[150] After publication, critics focused on the point that the paper was published in an Asian and not a Western scientific journal. Even ignoring this racism, the scientific facts do not change. Fred Singer has documented how the journals *International Journal of Climatology*, *Geophysical Research Letters* and *EOS* buried his research work because it did not fit the popular paradigm.[151]

Dispassionate indifference

For the first 300 years after the Royal Society was founded, they adopted a position of aloofness from political debates refusing to become embroiled in the political controversies of the day. In fact,

146: Mcintyre, S. and McKitrick, R. 2003: Corrections to the Mann *et al.* (1998) proxy data base and Northern Hemisphere average temperature series. *Energy and Environment* 14: 751-771

147: McIntyre, S. and McKitrick, R. 2005: Hockey sticks, principal components, and spurious significance. *Geophys. Res. Lett.* 32: L03710, doi:10.1029/2004GL021750

148: Montford, Andrew 2010: *The Hockey Stick Illusion*. Stacey International

149: McKitrick, Ross 2015: The hockey stick: a retrospective. In: *Climate change: The facts* (Ed: Alan Moran). Institute of Public Affairs, 201-211

150: Lindzen, R. S and Choi, Y. S. 2011: On the observational determination of climate sensitivity and its implications. *Asian Pacific Jour Atmos. Sci.* 47: 377-390

151: http://americanthinker.com/articles/2015/08/peer_review_is_not_what_its_cracked_up_to_be.html

their journal *Philosophical Transactions of the Royal Society* carried a notice:

> *It is neither necessary nor desirable for the Society to give an official ruling on scientific issues, for these are settled far more conclusively in the laboratory than in the committee room.*

This all changed in the 1960s and the Society gradually became increasingly involved in politics and policy. The last two presidents were especially active and it was only a revolt by Fellows that changed matters slightly. The Society now carries a damaged reputation because of its engagement in political controversies. The Royal Society is not alone. Almost all professional societies are involved in political lobbying, policy formulation and advocacy. Many learned societies, including the Royal Society, are partially funded by government and desperately need to sing from the same hymn sheet as the government otherwise funds will dry up and the society will become extinct.

Those societies to which I belong also have internal ructions when the executives and even employees untrained in science act against the wishes of members by taking a position on political controversies of the day. The same has happened elsewhere. The American Physical Society played the political game and, in 2010, revised its statement on climate policy. A scathing letter of resignation by one of its most eminent and senior members, Hal Lewis (University of California), commented on the popular paradigm of human-induced climate change:

> *It is the greatest and most successful pseudoscientific fraud I have seen in my long life as a physicist.*

My criticism of the climate industry is that much of the data (e.g. temperature and CO_2 measurements and corrections) is contentious, that computers have to be tortured to obtain the pre-ordained result, that computer codes are not freely available, that neither the data nor the conclusions are in accord with what we know from the present and the past, that publication is within a closed system that challenges the veracity of the peer-review system, that financial interests are not declared, that the financial rewards for publishing a scary scenario are tempting and that the process of refutation seems to have been overlooked.

The climate industry now needs to show it is dispassionate, has no pre-ordained conclusions, is unrelated to green left environmental politics, is driven by intellectual curiosity and can accept and critically analyse competing theories.

The quickest way to get the attention of the climate industry is to stop funding research that has a pre-ordained conclusion to show that human-emissions of CO_2 drive climate change and fund research on the long history of climate on planet Earth and research that deals with the possibility of natural global cooling. Let's follow the money and see the rats leave the sinking ship.

In the 1970s, the same people who were telling us we were going to freeze then started to tell us we will fry and die (e.g. Stephen Schneider). They exaggerated then and exaggerate now. Deceit is the one common thread that unites climate "science", as their own words show.[152] They left the sinking ship of global cooling and followed the money to the global warming ship. The scientific data is unchanged.

In areas of science where much of the debate by experts is obscure to the layman and relies on truthfulness (including uncertainties), the climate "science" community has let down the side and operated as green left environmental activists for certain policies by

---

152: "We need to get some broad-based support to capture the public's imagination. That, of course, entails getting loads of media coverage. So we have to offer up scary scenarios, make simplified, dramatic statements and make little mention of any doubts we may have … Each of us has to decide the right balance between being effective and being honest." Stephen Schneider, October 1989 interview for *Discover* magazine.

singing from the same tune as the funder of their "science" (i.e. governments). As Judith Curry[153] gently states:

> *In the climate change problem, it seems that often one's sense of social
> justice trumps a realistic characterisation of the problems and uncer-
> tainties surrounding the science and the proposed solutions.*

I am not so gentle. The global warming scare is the biggest scientific fraud in the 2,000-year history of science.

The rise of anti-intellectualism

With climate "science", not only are there huge amounts of research
funds sloshing around but also the economic implications of a carbon trading scheme, "Carbon Tax" and the transfer of wealth have
brought together all sorts of disparate and desperate groups. Trillions of dollars have been spent on the global warming scare. Science funding opportunities, the rise of the Green Party and financial
opportunities in the carbon market place have combined to produce
an almost perfect storm of quasi-religious hysteria with all the hypocrisy that one associates with a fundamentalist religion.

Fear of the environmental consequences of ever-expanding
economies combined with the deterioration in science education
encourages people to think of simple solutions to problems (or
non-problems). This is amplified by a lazy media that exploits fears
to increase sales and can't really be bothered to get to the bottom
of a story. This is not helped by journalists who want an instant
scary story, who are untrained in science and who are green left
environmental activists using journalism to promote their ideology.
In journalism, facts are sacrosanct. A little bit of basic knowledge
also helps. Do a test. Ask a green left environmental activist journalist how much of that really dangerous gas CO_2 is in their marble
kitchen bench top?

153: Curry, J. 2014 and 2015: *Ethics of climate expertise* and *Conflicts of interest in
climate science*, http://www.judithcurry.com

There has been a great decrease in science ethics. We now have senior scientists seeking to suppress publication of research that undermines their beliefs by setting up a taxpayer-funded web site to criticise an alternative view. We have senior scientists making statements that the "science is settled" on human-induced climate change regardless of contrary findings of others. Why would the Pope support those with dubious ethics? We now have previously well-regarded institutions discarding scientists or refusing to appoint those whose work is against popular opinion.[154]

Some university staff and students even agitate to stop the appointment of the sceptical environmentalist, Bjorn Lomborg.[155] If the University of Western Australia and Flinders University are unable to cope with a diversity of opinions, then clearly the wrong people have been appointed to and are running taxpayer-funded universities. Although Lomborg claims he's not a "climate change denier", which his writings show,[156] he demands that his opponents stop misrepresenting his views.[157] His critics claim that he wants to establish a climate change denial centre. One would have thought that scepticism and a breadth of competing ideas was healthy in a university.

If Flinders University caves in to arguments based on emotion and not logic and reason, then it is not protecting its reputation. The National Tertiary Education Union, the Flinders University Student Association and 6,662 students, staff and alumni who signed

154: Bjorn Lomborg: University of WA scraps research centre, *The Australian*, 8 May 2015

155: Students and staff warn of angry backlash if 'sceptical environmentalist' Bjorn Lomborg sets up research centre at Flinders University. *Socialist Moaning Herald*, 24 July 2105

156: Lomborg, B. 2001: *The skeptical environmentalist: Measuring the real state of the world*. Cambridge University Press

157: Bjorn Lomborg: I'm not a climate-change denier. *The Australian*, 26 July 2015

an open letter of rejection of the centre[158] are showing that the university's reputation would be based on anti-science, anti-intellectual, anti-democratic, anti-freedom of speech and anti-capitalism[159].

Many Australian universities, their staff unions and students associations have already been infiltrated by green left environmental activists and many universities have now become non-rigorous politicised third rate institutes. It was little different 40 years ago.[160] Who are these ill-educated undergraduate students to determine research directions? The age of open transparent science may be coming to an end and we are now seeing the end result of decades of dumbing down of the school and university education system. If you don't believe me, ask a private sector employer.

The history and evolution of planet Earth has been ignored with the current popular catastrophist story of human-induced climate change. If large bodies of evidence and history are ignored, then this provides an unbalanced, misleading and deceptive view of global climate. If scientists ignore integrated interdisciplinary empirical evidence, then they have politicised science to gain government favours and they are operating fraudulently. If scientists ignore history, then they do so at their peril.

Why is it that I have little confidence in the data, the methods, conclusions and science communication of scientists in the climate industry and of those who promote the doom-and-gloom story? It is a long list: (a) Climategate, (b) ignoring past climate changes, the rates of past changes and the natural variability of climate, (c) "adjustment" of primary data to yield the required result, (d) corruption of the peer-review process, (e) exclusion of contrary views by

---

158: *The Guardian*, 24 September 2015

159: We are reminded of the quote from Bertrand Russell's *Sceptical Essays*: "We are faced with the paradoxical fact that education has become one of the chief obstacles to intelligence and freedom of thought."

160: Greg Sheridan, 2015: *When we were young and foolish*. Allen & Unwin

eminent scientists, (f) corruption of both the temperature record and the CO_2 record, (g) non-correlation of CO_2 with temperature on all time scales, (h) conversion of science and free independent inquiry into political advocacy, (i) dampening or omissions of the validated record of the Minoan, Roman and Medieval Warmings, (j) creation of the "hockey stick" *ex nihilo*, (k) demonising of dissent, (l) denial that planet Earth changes by other forces far larger than anything humans can create, (m) failure of computer-generated models, (n) massive vested interests promoting the certainty of a human-induced catastrophe coupled with fraud and hypocrisy and the lack of caution and reserve in making public statements about new scientific findings. This is the thread of deceit that unites all green left environmental activists.

The gains made in the Renaissance and Enlightenment have been lost over the last two decades. Any system that allows questioning of beliefs is an enlightened system yet the climate industry is doing everything to stop the questioning of the basics of climate "science". Such behaviour as outlined is not that of scientists but of paid political thugs. Much of the political and media pressure comes from full-time climate advocates paid to misinform. And who will benefit? The taxpayer, of course, will pay. Banks, traders, alternative energy companies and investors, brokers and insurance companies are lining up in the shadowy world of totally opaque carbon trading to make astronomical profits.

No one has been able to accurately determine why climates changed in the past. We have some pretty good ideas and these we argue about as part of healthy scepticism in science. By the same token, no one can describe future climate. We have some pretty good ideas that should be debated. They are not. The only certainties are that a decarbonising of the economy will be a retreat from industrial progress and will increase global hunger as energy becomes more expensive. It may be that future energy will be rationed by self-interested cabals.

No amount of profits made from trading will remove CO_2 from the atmosphere. Despite the rhetoric, financial organisations are not interested in the planet. They are interested in profit. And rightfully so. It does not appear that we have learned very much from previous financial crises. The amount of funds transferred to bank profits means less money for genuine environmental programs. Government proposals for a green future are, in reality, proposals for opening the floodgates for the laundering of your tax money.

Science has always self-corrected. It takes time because many of those advocates of an incorrect theory must finish their working life and give up positions of power before such change can take place. Maybe the time of self-correction has ended because billions of dollars are sloshing around for climate "research" to give the pre-ordained conclusion that we are all going to fry and die and it's all our fault. Why bother funding anymore? We know the answer. Western governments have uncritically and dogmatically embraced human-induced global warming as an excuse for increasing taxation, redistributing wealth, eroding freedoms, constraining liberal thinking processes and maintaining power by doing deals with groups allegedly concerned about the environment.

The politicisation of science and the grubby Climategate affair is a weakening of science. We can only feed ourselves because of science. It is science that gives us longer and better lives. When the next inevitable pandemic hits humans, we are not going to solve problems by denigrating those with an opposing view and claiming consensus. Governments may be so embittered by the human-induced global warming scam that there may be no funding for disciplines such as medical science. In a pandemic, we would need science, independent thinking, creativity and for scientists to tread where no one else has been. That is the only way to solve problems but these very problem-solving processes from the Enlightenment are being undermined and destroyed by the climate industry.

We frail humans (including scientists) commonly fall prey to fads, fashions, frauds and fools because we ignore the past. Some people actually think that we humans may be important in the scheme of planetary matters yet seem unaware that one volcano could ruin their whole day and a bacterial pandemic could see us off forever.

We cannot ignore social, political, economic and Earth history.

CLAIMS OF THE CLIMATE CATASTROPHISTS

Unprecedented temperature changes over the last 100 years

WRONG. The claim that the Earth's atmosphere at the surface has warmed by 0.85°C over the last 100 years is unprecedented and is not in accord with evidence. The Central England Temperature Record shows that between 1693 and 1733 AD temperature rose at a rate of 4°C per century. During the Roman and Medieval times it was warmer than now and the rate of warming at times was higher than in the 20ᵗʰ century.[161,162]

Furthermore, global temperatures have increased some 3 to 8°C since the zenith of the last glaciation, global temperatures have a variability of about ±3.5°C from the long term mean[163], in the 21ˢᵗ century we are about 1°C above the long term mean but, before you go outside and slash your wrists, in the Medieval Warming the temperature was 2.5°C above the long term mean. There is nothing

161: Christiansen, B. and Ljungqvist, F. C. 2012: The extra-tropical Northern Hemisphere temperature in the last two millennia: reconstructions of low frequency variability. *Clim. Past* 8: 765-786

162: Ljungqvist, F. C. *et al.* 2012: Northern Hemisphere temperature patterns in the last 12 Centuries. *Clim. Past* 6: 227-249

163: Jouzel, J. *et al.* 2007: Orbital and millennial Antarctic climate variability over the past 800,000 years. *Science* 317: 793-796

special about the 21st century temperature or temperature increase over the last 100 years. It is just the normal variability that planet Earth enjoys.

Warmest year on record was 2014

WRONG. The record, and it is a poor one at that, only goes back to 1850 (HadCRUT4, thermometer), 1880 (NASA, GISS, NCDC, thermometer) or 1979 (RSS and UAH, satellite). The thermometer records are "adjusted" after which it appears that 2014 was the warmest year. The satellite data shows that 2014 was not the warmest year. The normal climate "science" trick for a time period is to start the record at a period of low temperature (or even "adjust" the temperature downwards) and close the time record at a warm period (or "adjust" the measurement upwards).

There has been no global or regional temperature increase for more than 18 years despite the alleged spike in temperature for 2014. The long-term record using proxies shows a different story. The Holocene maximum (9,000 to 5,000 years ago) and Egyptian Old Kingdom, Minoan, Roman and Medieval periods were warmer than now. It is often suggested that the frequency of high temperature extremes is increasing. What is not mentioned is that there have been just about as many cold weather records as hot weather records. Furthermore, in the 1930s there were many periods of extreme temperature although many of these have been "adjusted" hence expunged from the record. The weather record is somewhat like sporting records. Somewhere sometime in the world a record is broken.

I heard the other day that July 2015 was the hottest Northern Hemisphere summer month on record and that 2015 looks like it will be the hottest year in the 136 years of data collection. The imputation was that this was due to human activities and that global warming continues unabated. Something odd is happening in the North Pacific Ocean. Warm water is not where it should be which may be related to the delayed 2014-2015 El Niño event. El Niños are followed by La Niñas and, while 2016 may also be slighter warmer, subsequent years may be somewhat cooler.[164] This changes sea surface temperature, weather patterns and ecosystems. Records since 1905 show that this has not happened before. Such events influence weather, not climate. However, this 2015 warming is not seen in the satellite data and the Southern Hemisphere temperature record, despite "adjustments", shows that July 2015 was a colder than normal winter.

It appears that the closer we are to the 21st session of the Conference of the Parties (COP-21) of the United Nations Framework Convention on Climate Change (UNFCCC) to be held in Paris from 30 November to 11 December 2015 the hotter the planet had become.

The ice sheets and glaciers are melting and sea ice is contracting

WRONG. Some glaciers are expanding, others are retreating. This is normal and no cause for alarm and certainly can't be used to give scary predictions. Ice sheets and sea ice show a long history of expansion and contraction, again no cause for alarm.[165] The amplitude of seasonal variation of sea ice in both the Arctic and Antarctic dwarfs the changes in sea ice that are meant to show catastrophic global warming. Furthermore, the extent of global sea ice shows remarkably little change over the past 35 years[166] and even the IPCC shows that there is little or no warming in Antarctica.

President Obama and Secretary of State John Kerry visited Glacier Bay in Alaska in late August 2015. They pointed to the receding glaciers as evidence that humans are the cause of dangerous catastrophic climate change. A bit of history rather than hype would have

---

164: David Whitehouse, Global Warming Policy Forum, 6 September 2015
165: Braithwaite, R. L. 2002: Glacier mass balance: The first 50 years of international monitoring. *Prog. Phys. Geog.* 26: 75-95
166: http://arctic.atmos.uiuc.edu/cryosphere/

been more honest. The glacier started to retreat in 1750 AD, when Captain George Vancouver visited in 1794 the ice filled most of the bay and had only retreated a few kilometres. When the founder of the Sierra Club, John Muir, visited Glacier Gay in 1879, he found that the ice had retreated 50 km from its 1750 position. In 1900, it was almost free of ice. This shows that there is local climate change and that it is unrelated to human emissions of CO_2 since the Industrial Revolution. Maybe it is related to the 300 years of warming we have enjoyed since the Maunder Minimum.

Sea level is rising rapidly, Pacific islands, the Maldives and coastal cities will be inundated

WRONG. The theory is that as the planet's atmosphere warms, the oceans will warm and expand. Air has a low heat capacity, water has a high heat capacity. If the oceans cool, the air temperature decreases. If the oceans warm, the air temperature rises. Atmospheric temperature does not drive warming or cooling, it is the oceans. According to the GRACE gravitational recovery satellites,[167] sea level actually fell from 2003 to 2008. The ENVISAT[168] shows that sea level rose at a mean rate of 3.3cm per century.

Just to put these estimates into perspective, the inter-calibration errors between the series of laser altimetry satellites from the official sea level record is greater than the sea level change they purport to show. Tide gauges and benchmarks show very little sea level rise. Some of the purported sea level rise has actually been shown to be due to the subsidence of the measuring station. The ARGO bathy-thermographs[169] show that during the first 11 years of measurement (2004-2014), the oceans to nearly two kilometres depth warmed at a rate of 0.2°C per century.

167: http://www.csr.utexas.edu/grace/
168: https://earth.esa.int/web/guest/missions/esa-operational-eo-missions/envisat
169: http://www.oco.noaa.gov/xBTsSOOPS.html

Rainfall extreme weather events have increased

WRONG. The IPCC in its 2012 and 2013 reports show that there is no evidence of increased extreme rainfall events due to human activity. The 250 years of rainfall records in Britain shows little change as does the 100-year record in the US. Furthermore, deaths from extreme weather are currently at an all time low despite increases in the atmospheric CO_2 and population.[170]

Global warming will bring more drought

In Australia, global warming alarmists panicked Labor governments into creating subsidised wind and solar power and building massive desalination plants during a drought. Four desalination plants built by former state Labor governments in Australia that have since been mothballed will cost taxpayers nearly $1 billion in 2015.[171] It will be more in 2016 and there are no plans for their use. They were built at times when green left environmental activists were screeching about how Australian cities would run out of water.

Maybe Labor politicians did not learn the second verse of Dorothea Mackellar's *My Country* when they were at school:

> *I love a sunburnt country,*
> *A land of sweeping plains,*
> *Of ragged mountain ranges,*
> *Of droughts and flooding rains.*
> *I love her far horizons,*
> *I love her jewel-sea,*
> *Her beauty and her terror –*
> *A wide brown land for me!*

Australia has droughts. It is a brown country. It also has flooding rains. In 1911, Dorothea Mackellar was correct. She is still correct.

170: Global Warming Policy Foundation 2014
171: *The Weekend Australian*, 12-13 September 2015

Drought has been and will continue to be major feature of Australia. Green left environmental activists, the media and city-based politicians are unaware of drought because it does not affect them directly. Maybe they should travel in outback South Australia and see the abandoned homesteads and villages where dreams were broken in the 1880-1886, 1888 and 1895-1903 droughts.

There have been ten major droughts in Australia over the last 150 years.[172] Some droughts lasted a decade, some are in clusters and all had a severe effect on rural Australia. Primary production decreased, banks foreclosed businesses, unemployment rose, rural towns died, farmers committed suicide and government assistance was too little and too late. City life blissfully continued unaware of the tragedy on their doorstep and the only impact was that food prices of selected products increased slightly. Many droughts were broken by "flooding rains."

The Australian Bureau of Statistics[173] and Bureau of Meteorology[174] document the major Australian droughts (1864-1866, 1868, 1880-1886, 1888, 1895-1903, 1911-1916, 1918-1920, 1939-1945, 1958-1968, 1982-1983, 1999-2009, 2012-2015), less severe droughts (1922-1923, 1926-1929, 1933-1938, 1946-1949, 1951-1952, 1970-1973, 1976) and droughts where most Australian live (i.e. SE Australia; 1888, 1902, 1914-1915, 1940-1941, 1944-1945, 1967-1968, 1972-1973, 1982-1983, 1991-1995, 2002-2003, 2006-2007).

The three biggest droughts were 1895-1903, 1939-1945 and 2002-2007.[175] However, rather than take advice from government departments that deal with historical information, the last big drought led

172: Couglan, M. J. 1986: *Drought in Australia. Natural Disasters in Australia.* Australian Academy of Technological Sciences and Engineering
173: http://www.abs.gov.au/AUSSTATS/abs@.nsf/lookup/1301.0Features%20Article151988
174: http://www.bom.gov.au/climate/drought/
175: http://www.australia.gov.au/about-australia/australian-story/natural-disasters

to political panic aided and abetted by green left city-based environmental activists and cheered along by the hysterical ABC and Fairfax media networks. Australia's most unsuccessful climate forecaster (Tim Flannery) was able to state without critical questioning from the media:

> *So even the rain that falls isn't actually going to fill our dams and our river systems, and that's a real worry for people in the bush ... I think there is a fair chance Perth will be the 21st century's first ghost metropolis ... Perth is facing the possibility of a catastrophic failure of the city's water supply ... I'm personally more worried about Sydney than Perth. Where does Sydney go for more water?*

This was a great scare campaign and neither the media nor Flannery bothered to tell people that drought is normal, that there have been many long hard droughts in Australia before and that previous droughts occurred before Australia was industrialised and emitting plant food (CO_2) from transport and industry. No one in the mainstream electronic or print challenged Flannery's scary claims or qualifications. Commentators who questioned Flannery's catastrophist claims were ignored. Flannery has degree in English Literature and a PhD on tree kangaroos hence is the perfect media go-to person for expert opinion on drought. In 2007, he opined: "In Adelaide, Sydney and Brisbane, water supplies are so low they need desalinated water urgently, possibly in as little as 18 months."

The Queensland Labor Premier spent more than $1.2 billion of taxpayers' money on a desalination plant on the basis that the lower than usual rainfall would continue. It didn't. The expensive desalination plant has been mothballed and Brisbane has since been flooded twice. Not to be outdone, the Victorian Labor government spent $2 billion on a desalination plant and, at the time of its opening, catchment dams were more than 80% full. The plant was mothballed. Sydney also spent horrendous amounts of money on a desalination plant, since mothballed. It has produced no water since 2012.

In 2008 Flannery stated: "The water problem is so severe for Adelaide that it may run out of water by early 2009."

It didn't. At times, Adelaide water is character-building and, when mixed with sand and aggregate, can make excellent concrete. Adelaide also got into the act and spent $2.2 billion on a 100 Gigalitre desalination plant that has since been mothballed. For the poorest mainland state in Australia with the highest unemployment to spend billions on a white elephant shows how green ideology has infiltrated politics and, no matter what the cost, the public pays.

The southeastern states of Australia recently spent more than $10 billion building desalination plants. All were mothballed. This was a shocking waste of public funds resulting directly from a green left environmental activist scare campaign. Politicians, media, academics and green left environmental activists were not held to account, they all fell for the global warming scare and have since moved on to try to scare us with other aspects of their predicted global warming disaster. Why should we ever take notice again of green left environmental activists? No person or company in the productive part of the economy has ever made such horrendous mistakes based on propaganda and without a full due diligence process.

To make matters worse, despite failed predictions, Flannery and his fools still get airtime with a compliant media on any issue that concerns the weather, climate and the environment. I guess that this is one of the hallmarks of the green left environmental movement, one just has to get airtime and every activist is an expert.

In Perth, it was different. There had been a slight decline in rainfall (especially in winter) over the last 40 years in southwestern Western Australia.[176] Such cycles are common all over Australia. Furthermore, the slightly lower runoff increased salinity of waters captured in dams (e.g. Wellington Dam) and much dam water became unsuit-

---

176: www.cawcr.gov.au

able for drinking. Perth was a rapidly expanding metropolis, much domestic and industrial water was from a shallow aquifer and new water supplies were needed.

A desalination plant was built near Kwinana (Perth Seawater Desalination Plant) and supplied Perth with 17% of its water needs. The second and bigger plant was built further south at Binningup (Southern Seawater Desalination Plant).[177] Some 50% of Perth's water is now from desalination plants and the energy for these plants derives from fossil fuel (natural gas). Water Corporation WA also purchases the entire output from the 10 MW solar plant at Greenough and from the 55 MW wind plant at Mumbida and claims that the Binningup plant is supposedly "carbon neutral" (except of course at night when the wind does not blow).

Perth residents did not suffer from climate change. They had a population explosion. Residents also did their bit and changed from well-watered green lawns to native gardens that needed less water. The additional costs have been absorbed into the Water Corporation WA costs and sandgropers are just happy to have water. The Pope would approve of the community working together and changing from water-hungry lawns to native flora and would also approve of using desalinated water rather than runoff and ground water.

The normal suspects were up in arms about building a desalination plant at Kwinana. They claimed that the sea grass meadows off Kwinana were fragile, and that there would be an increase in salinity in the Indian Ocean. The sea grass was going to die, this would create an ecological collapse and that everything off the coast of Western Australia would die. The sea grass did not die. A simple mass balance calculation would show that there would be no measurable increase in salinity in the Indian Ocean off Kwinana as dilution is

---

177: www.watercorporation.com.au

effectively infinite. As usual, science and engineering solved a problem. As usual, the green left environmental activists were hopelessly wrong and just moved onto the next perceived disaster.

Maybe the best solution is to give Tasmania to the greens so they can live their pure moral life without the evils of modern society. Let them try to feed themselves, cook, stay warm and create employment and the rest of continental Australia can be a green-free wide brown land where people can just get on with the job of living life.

Clean desalinated water costs a huge amount of money. It is out of reach for the developing world. There are many other ways of producing clean water for humans.

Increasing human emissions of CO_2 are affecting global temperature

WRONG. Although atmospheric CO_2 has risen by 40% since the Industrial Revolution, this does not mean that humans have affected global temperature. As an example, if the atmosphere comprised 85,000 molecules, then the total CO_2 emissions added annually would be 33 molecules with the human contribution being one molecule. It has yet to be shown that this one CO_2 molecule from human activities drives climate change and the 32 other molecules emitted from natural sources don't. It's rather like having the Melbourne Cricket Ground packed with 85,000 spectators and claiming that one bellowing fan amongst 85,000 has the ability to change the course of the game.[178] The carbon cycle, the oceans, plants and microorganisms don't recognise that a molecule of CO_2 derived from human emissions is different from a molecule exhaled from the oceans. Plans to bury this one molecule of CO_2 could only have been devised in a funny farm or in a generous embrace with Bacchus. Or is there some other devious reason?

Let's get this into perspective: CO_2 is a trace gas in the atmo-

---

178: This is a claim that could only be made by Collingwood supporters.

sphere, the rare inert gas argon is far more abundant. If we humans add traces of a trace gas to the atmosphere then we are making very little change to the atmosphere, especially as a CO_2 molecule has a life of about five years in the atmosphere before it is naturally sequestered. Some 750 million years ago (Ma), the atmospheric CO_2 content was about 1,000 times the current level. The planet didn't fry, it actually had an ice age.

When there was an explosion of life on Earth about 550 million years ago, the planet had up to 25 times as much atmospheric CO_2 as at present. In times when vegetation evolved and grew rapidly (e.g. Jurassic), the atmospheric CO_2 content was up to 15 times the present amount.[179] Are we doomed if CO_2 doubles? The past tells us that it will be business as usual with cyclical climates but computer model predictions show a 1.4 to 4.5°C warming if CO_2 doubles. Over the last 25 years, these models also have exaggerated the effects of warming and the IPCC has now given an estimate of 3°C warming if the atmospheric CO_2 content doubles. However, the more CO_2 in the atmosphere the more natural sequestration. We fail to hear about the benefits of additional atmospheric CO_2. For example, the net primary productivity of plants has increased worldwide by 2% due, in part, to a slight increase in atmospheric CO_2.

Sophisticated computer models can be used to predict future climate

WRONG. The models have been tried and tested. They have failed. They could not even confirm past climates by running backwards without substantial retuning. The real test is whether these models can predict future climate in say 100 years time. There are over 100 variants of climate models yet they did not even predict a period of no warming for at least 18 years. If they couldn't even get this right, then no matter how much tampering and adjustment, we can safely

179: Petit , J. *et al*. 1999: Climate and atmospheric history of the past 420,000 years from the Vostok ice core, Antarctica. *Nature* 399: 429-436

conclude that they can't predict what will happen in 100 years time. In 1990, the IPCC predicted that the rate of global warming would be twice what occurred. In 2007, the IPCC predicted that in the decade following 2005, there would be significant warming. There wasn't even any warming in the decade following 2005.

In 2013, the IPCC were at it again and predicted short-term warming. We're still waiting. IPCC models predicted that there would be an atmospheric hot spot above the tropics at an altitude of about 10 km. Despite the release of about 30 million weather balloons since 1950, the modelled hot spot cunningly hid from every single one of these balloons.

The climate "science" models made two predictions. The first was about temperature. These were grossly exaggerated and wrong. The second was about a hot spot. This was wrong. Both were shown to be wrong as a result of measurement.

It is safe to conclude that pretty well everything you hear, read or see in the popular media about climate change is wrong, exaggerated or made up. And if models are used, you can be certain that it is wrong.

CLIMATE MODELS AND PREDICTIONS

The Encyclical contains the statement that 2015 was the "*last effective opportunity to negotiate arrangements that keep human-induced warming below 2 degrees*". How does the Pope think that just by twiddling the dials we can control global temperature? Hunters in the jungles of Sarawak might have a different idea of the ideal temperature to people in Saudi Arabia, Siberia or Scotland. I'm sure those at high latitudes would prefer a slight warming of 2°C. After all, they go to warmer climates for their holidays. For me, the ideal dry climate temperature is 28 to 38°C.

Is the Pope aware that the human body works best at an air temperature of 19°C yet the UK average atmospheric temperature (for 2014) was less than 10°C. Would not the average Siberian enjoy a 2°C temperature rise? This would create an agricultural boom in Russia, Canada, Greenland, Scotland, Faroe Islands, Norway, Sweden, Finland and Iceland. Just think how human emissions of CO_2 would be reduced if the global temperature was increased 2°C. Less household heating in the cool parts of the Northern Hemisphere would be required.

And why 2°C. This is just a number plucked out of the air. During previous times, the planet's temperature has changed by ±3.5°C. People didn't die like flies in warmer times and premature death was more common in colder times. The Pope's statement is just scaremongering. We have seen it many times. For example, the French Foreign Minister Laurent Fabius stated to the US Secretary of State John Kerry on 13 May 2014 that:

> … *we have 500 days to avoid climate chaos.*

At the time of Fabius' comments, the UN had just scheduled a climate summit for Paris in late November-early December 2015, some 500 days from the Fabius' prediction.

Climate conferences always attract predictions from those with a self-interest and all sorts of deluded people come out of the woodwork and make predictions. Not one previous prediction has been correct.[180] World leaders met in Copenhagen in late 2009 to try to twiddle the dials to stop global warming when outside there was thick snow and it was bitterly cold. The leader of the greens in Canada wrote in 2009 that *"we have hours to act to avert a slow-motion tsunami that could destroy civilisation as we know it"* and

> *Earth has a long time. Humanity does not. We need to act urgently.*
> *We no longer have decades; we have hours. We mark that in Earth*
> *Hour on Saturday.*

180: http://dailycaller.com/2015/05/04/25-years-of-predicting-the-global-warming-tipping-point/

This reads like the rantings of someone emotionally unbalanced.

In 1982 the UN was already telling the world that it only had a decade to solve global warming or face the consequences. Nature did the job and there has now been no global warming for more than 18 years. We can thank nature for showing us how little we really understand about Earth systems. In 1989, Noel Brown, a senior UN environmental official, claimed that:

> ... entire nations could be wiped off the face of the Earth by rising sea levels if global warming is not reversed by the year 2000.

We've had a few new nations appear rather than disappear since 1989 (e.g. South Sudan) and the prediction was wrong in 1989, wrong in 2000 and it is still wrong. The UN's IPPC former head, Rajendra Pachauri, stated in 2007:

> If there's no action before 2012, that's too late. What we do in the next two to three years will determine our future. This is the defining moment.

All that has happened since 2007 is that temperature has not increased. Pachauri's future was certainly determined in a defining moment. He resigned in 2015 as the IPCC head after allegations he sexually harassed many female employees. Again, in 2007, the UN reported that global emissions must peak by 2015 for the world to have any chance of limiting the expected temperature rise of 2°C. The temperature has not risen, the temperature has not fallen and CO_2 emissions continue to rise because more and more people in the Orient are enjoying a better standard of living. How long can the UN cry wolf?

The IPCC is embarrassed. It has now had to admit that the climate models built by their so-called climate scientists cannot predict the temperature that is measured. They meekly state:

> For the period from 1998 to 2012, 111 of the 114 available climate-

model simulations show a surface warming trend larger than the observations.

This is the coward's way of saying that they were wrong.[181]

Obama made a campaign promise to *"slow the rise of the oceans"* and Obama was advised in 2012 by the UN Foundations President (Tim Wirth) that his second term was the *"last window of opportunity"* to reduce the use of fossil fuels in the US. Obama is trying to do exactly that. Wirth advised the President that it's the:

last chance we have to get anything approaching 2 degrees Centigrade

and

if we don't do it now, we are committing the world to a drastically different place.

Obama attracts unsolicited advice and climate activist James Hansen warned in 2009 that Obama only *"has four years to save the Earth"*. That was six years ago. Not only have I missed trains and planes, but it appears that I also missed the end of the Earth.

The UK has its share of doomsdayers. Prime Minister Gordon Brown in 2009 stated that there were only *"50 days to save the world from global warming"* and, following on from his hysterical prediction, claimed that there was "no Plan B". At that time there had been 12 years without warming. In July 2009, Prince Charles said that there would be "irretrievable climate and ecosystem collapse, and all that goes with it" and that we had 96 months to save the planet. I don't know how he arrived at that figure but 72 of those months have been and gone with no change to the climate and ecosystem.

Green left environmental activist journalist George Monbiot, the master of failed predictions, wrote in 2002[182] that:

181: http://www.economist.com/news/science-and-technology/21598610-slowdown-rising-temperatures-over-past-15-years-goes-being/
182: http://www.theguardian.com/uk/2002/dec/24/christmas.famine

Within as little as 10 years, the world will be faced with a choice: arable farming either continues to feed the world's animals or continues to feed the world's people. It cannot do both.

and

The impending crisis will be accelerated by the depletion of both phosphate fertiliser and the water used to grow crops.

Sorry George, the world's people 13 years later still get a feed and neither phosphate fertiliser nor water have run out. According to the UN, at the time of his 2002 prediction, about 930 million people were undernourished. By 2014 the number was 805 million despite a population increase from 6.28 to 7.08 billion people.[183] No wonder George is known affectionately in England as Moonbat. It's not hard to see why.

The UK Met Office wins the gold medal for failed predictions. They undertook an investigation from 2004 to 2014 on the effects of future climate change. The Met Office's computer models assume that a rise in atmospheric CO_2 is driving climate change. Human emissions of CO_2 are rising, atmospheric CO_2 is rising (which may or may not be related to human emissions of CO_2) and the average global temperature has not done what the computers told it to do. For nearly two decades, global temperature has not risen. Despite this, the Met Office has continued to tout scary scenarios for the future.

The Met Office receives £220 million *per annum* from the taxpayer and yet they want bigger and better computers in order to make more inaccurate predictions more quickly. According to the Met's chief scientist, their £33 million supercomputer is just not good enough for accurate predictions and they now want a £97 million supercomputer. One prediction was that at least three of the years after 2009 would be hotter than the El Niño year of 1998 (when it is hotter anyway). And what happened? The Met Office claimed

183: http://www.faq.org/hunger/en/

that 2010 and 2014 were hotter than 1998 but this was only after "adjusting" the measured and published sea and air surface temperatures upwards. There was no scientific justification as to why this was valid. Furthermore, satellite measurements showed that 1998 was a hot year and 2010 and 2014 were nowhere near what were previous hot years. The Met Office conveniently ignored the more accurate satellite data.

Another Met Office claim was that there would be more heat waves like the 2003 heat wave when an extra 15,000 people checked out early. However, at the time, the same meteorologists stated that the 2003 heat wave was nothing to do with climate change; rather it was due to an unusual influx of hot air from the Sahara. Extreme weather events, such as abnormal rainfall were another Met Office scare scenario. In 2014 there was heavy rain and flooding but the Met Office's own records show that far more rain fell between December 1929 and January 1930.

The global warming poster child, Greenland, is always good for a scare. The Met office claimed that at some unspecified time in the future, there will be a melting of the Greenland ice cap, sea level will rise by six metres and coastal and estuarine cities will be engulfed. This is actually a correct prediction because, despite six major ice ages, ice has only been on Earth for less than 20% of time. Some time in the geological future when our current ice age stops, the polar ice will melt. I predict that this will be tens to hundreds of millions years away and will occur at five past on a Thursday. Maybe the Met Office was not aware of studies in Greenland that show since 1900 AD that the Greenland air temperature has not risen.

During this period from 2004 to 2014 when the climate doomsday book was being compiled, the Met Office was making additional predictions by playing £33 million computer games. It was predicted that 2007 would be the "*hottest year ever*". This prediction was just before global temperatures decreased by 0.7°C. They also predicted that

2007 would be *"drier than average"*. It was a very wet year with some of the worst floods in recorded history. Between 2008 and 2011, the UK was to experience *"warmer than average"* and *"hotter and drier"* summers, or so the computer said. These were some of the coolest and wettest summers on record, even though there was a short time in 2009 that was called the *"barbeque summer"*. In October 2010 as I was preparing to leave for a conference in London in December, I was terrified to learn that the Met Office computer predicted that winter would be *"two degrees warmer than average"*. I packed my normal winter gear and endured the coldest and snowiest December since records began in 1659 AD.

Here is a credulity test for you, dear reader. A November 2011 Met Office computer forecast claimed that global temperature would rise by as much as 0.5°C by 2017. What do you think? Someone in the Met Office agrees with you and, after only a year, this prediction was removed from the Met Office web site.

In March 2012, we learned that spring would again be *"drier than average"*. This prediction was made just before the wettest April ever. Not to be outdone, in November 2013 it was predicted that the UK winter would be *"drier than usual"*. Thousands of people breathing through snorkels disagreed as the UK had the wettest three months on record. The Met Office predicted a warmer, drier summer than average for 2015. The rain poured down in August and the summer was the 178th warmest since records began in 1659 AD. The English were left shivering in the rain.

If only the UK Met Office would predict warm sunny weather for an Ashes series then Australia might not get comprehensively thrashed because, London-to-a-brick, there would be torrential rain and play would have to be abandoned.

Climate "scientists" did not predict record crop yields. They did not predict record sea ice around Antarctica. They did not predict a rapid growth of Arctic sea ice. They did not predict fewer hurricanes and cyclones.[184,185] They did not predict colder European winters. They did not predict more snow in the Northern Hemisphere. They did not predict dam-filling rains in Australia. They did not predict a failure of the Great Barrier Reef to bleach. They did not predict that for more than 18 years there would be no warming. Why should the taxpayer fund or even take any notice whatsoever of predictions by proven failures.

POLITICS OF CLIMATE CHANGE AND ENERGY POLICY

In the media's hysterical embrace of the paragraphs dealing with climate change in the papal Encyclical, few journalists seem to have read the whole document. The green left environmental activists and the Pope are in favour of renewable energy.[186] However, renewable energy comes with leg irons called carbon trading, fuel poverty and unreliable expensive electricity. The Pope seems to be against carbon trading.[187] He is in good company.

The Pope does not explain market forces, seems to be against a free market, does not see that free market forces have solved many past environmental problems and seems to regard company or individual profit as something undesirable. Yet profits result in employment, payment of taxes and these taxes fund the needs of the poor and funds for bureaucrats, politicians, universities and institutions telling us that we must destroy the economy or we'll fry and die.

184: Pielke, R. 2012: Hurricanes and human choice. http://www.wsj.com/articles/SB10001424052970204840504578089413659452702
185: http://www.drroyspencer.com/2015/05/nearly-3500-days-since-major-hurricane-strike-despite-record-high-co2/
186: *Laudato Si'*, Paragraphs 26, 164, 165 and 179
187: *Laudato Si'*, Paragraph 190: "Here too, it should always be kept in mind that environmental protection cannot be assured solely on the basis of financial calculations of costs and benefits. The environment is one of those goods that cannot be adequately safeguarded or promoted by market forces."

The land of the free

Climate change and the EPA

President Obama described climate change as *"one of the key challenges of our time"* as he announced the first ever limits on US coal-fired electricity generation plants. And when did Obama make this announcement? On the hottest day of the 2015 summer. In 2014, coal provided 39% of electricity in the US[188] and Obama is seeking to have a 32% reduction in CO_2 emissions from coal-fired electricity generator plants. This Obama described as *"the single most important step America has ever taken in the fight against global climate change"*. Really! And what will this single most important step achieve? The climate has not changed for almost two decades. To gamble trillions of dollars of US economy on prevention of 0.01°C warming and stopping the rise of sea level by 0.3 mm (the same as the thickness of three sheets of paper) overshadows all reality.

On many matters, elected politicians have handed over matters to unelected bureaucrats to create and use regulations. The US Environmental Protection Agency's (EPA) Clean Power Plan (CPP) requires that states reduce their CO_2 emissions from electricity generation to 32% below the 2005 levels by 2030. This is equivalent to the CO_2 emissions in 1975. The US population has grown by 40 million since 1975. At least 12 states will have to reduce emissions by 40-48%. These states produce 50-90% of their electricity from coal and natural gas. In the coal-reliant states, costs will rise from 8-9c/kWh to 36-40c/kWh, equivalent to the high costs in Denmark and Germany. The US is efficient and competitive because of the low cost of energy. Millions of US jobs are dependent upon the low cost of energy.

188: http://www.eia.gov, US electricity generation: coal 39%, natural gas 27%, nuclear 19%, hydropower 6%, wind 4.4%, biomass 1.7%, petroleum 1%, solar 0.4%, geothermal 0.4% and other gases (e.g. hydrogen, methane) <1%

The social costs of increasing energy costs will be profound yet the EPA costs energy by calculating a *"social cost of carbon"* that arbitrarily inflates the costs of speculative climate change damage from using fossil fuels. The EPA is unelected. The elected members of Congress have rejected more than 700 climate bills and it is clear that the EPA is acting against the will of the people. The EPA keeps its questionable data and analysis secret and refuses to share them with Congress or state governors. It ignores the fact that while atmospheric CO_2 emissions have been increasing, global temperature has not been increasing for more than 18 years.

Economic and social pain will hit the poor and minority families the hardest just for a modelled increase in atmospheric temperature of 0.018°C in 85 years time. This number is hardly measurable. The EPA's CPP document was called *"The Final Rule"*, a wonderful title that has a cheery Stalinist ring to it. Is the EPA really interested in climate, society and the environment or is this a game of power in the absence of political leadership?

The EPA often justifies its existence by noting that corporations, who see profit as their goal rather than environmental protection, are ill-equipped to care for America's natural resources. When EPA crews tried to collect wastewater confined in a stable situation in the Gold King Mine (Durango, Colorado), they released 4.2 million litres of acidic, yellow discharge rich in arsenic, lead, cadmium, aluminium and copper into the Animas River (a tributary of the Colorado River) at 45,000 litres per hour. The EPA claimed responsibility for both the pollution and its slow response. The incompetent EPA individuals don't pay for the clean-up. This, of course, falls on the taxpayer. Again. If a corporation created such a mess, executives would go to gaol.

US coal and nuclear

At the very moment President Obama decided to destroy America's satanic coal industry and replace it with far more expensive and less efficient renewable energy, the use of coal elsewhere in the world increased because coal is cheap, reliable and efficient.[189] America is certainly making itself uncompetitive. The US is closing coal-fired electricity generators but not as quickly as India is building them. Some 66 nuclear power stations are currently under construction around the world.[190] Only four new nuclear power stations are being constructed in the USA, 30 in China, six in India. Around the world, there are another 150 nuclear power plants on the drawing board with only five in the US.

The last nuclear power station built in the USA was Watts Bar 1 in Tennessee in 1996. The US has 99 operating nuclear reactors across the US. About 20 nuclear plants are scheduled to close. It is now probably easier to build a nuclear power station in Iran than in the USA. The global coal renaissance is in the fast growing economies of East Asia, India and parts of Europe where the proportion of coal and nuclear are increasing in the energy mix. Over 1,200 coal-fired electricity generating plants are planned across 59 countries, with about three-quarters in China and India.[191]

Coal use around the world has grown four times faster than re-newables, for obvious reasons. Renewables demonstrably can't competitively deliver electricity when it is required. In Germany, coal will remain the largest electricity fuel source for decades. By 2020, India will have built about 2.5 times as much capacity as the US is about to lose.[192] Between 2010 and 2013, China added coal-generating capacity equivalent to half of all the coal-generating capacity in the USA. For seven consecutive years, China has built two 600 Mw coal-fired

189: *Investors Business Daily*, 7 August 2015

190: http://bloomberg.com/news/articles/2015-04-15/soon-it-may-be -easier-to-build-a-nuclear-plant-in-iran-than-in-the-u-s/

191: World Resources Institute, www.wri.org

192: http://www.bp.com/en/global/corporate/about-bp/energy-economics/statistical-review-of-world-energy.html

electricity generating plants each week. This will not stop, China has a lot of catching up to do from a very low base.

As the US uses less coal, the rest of the world uses more. And to really rub salt into the wounds, the US has more coal than any other country and, compared to the developing world, can mine coal more cheaply and more safely. Coal is mined in some of the more impoverished areas of the US where there are few factories and agricultural enterprises and provides a welcome economic injection with jobs and a flow-on effect. The US has over 300 years of coal with a value of trillions of dollars. The US is the coal equivalent of oil rich Saudi Arabia. Furthermore, US coal is good quality coal that can easily be cleaned to a better quality lower ash and lower sulphur product.

Obama's legacy

Normally in the second term of a US president, there is a push to leave a legacy. President Obama will leave with a dubious legacy. He will go down in history as the President who destroyed cheap electricity and competitive industry in the US and greatly raised the energy costs of the poor and minorities. These are the very people who mostly vote Democrat. This massively expensive gesture will cost hundreds of thousands of jobs, raise utility prices by as much as $1,000 per family and will reduce the GDP by 0.5% at a time when the US is barely growing. Some legacy!

Obama's single most important step to making the US uncompetitive, increasing unemployment and frightening away investment is his hostility to coal. Such statements on CO_2 and the coal industry can only be posturing by President Obama to give him dubious international hero status at the United Nations Framework Conference in Paris and to create a perception that, in his second term as President, he is leading the way on combating emissions and that he actually did something for the world environment. It is not about policy, it is about Obama's egotistical legacy. He will go down in history, for the wrong reasons and not for saving the world from alleged global warming. China and India are not reducing their coal consumption and CO_2 emissions. And why should they? Reduction in US coal consumption will do nothing to reduce global CO_2 emissions.

The EPA Administrator Gina McCarthy claimed that the estimated additional cost by 2030 for renewable energy will be $8.4 billion *per annum*. Government estimates are always low and it is not the government's money. It is money paid by the hard working taxpayer. The aim is to have 28% of the nation's electricity from renewable energy by 2028, a rise from the current 13%. The bottom line is the closure of coal-fired power stations, stopping the expansion of the use of natural gas and, of course, the appointment of 800 new regulators to add to the 15,000 currently in the EPA. Is this the American dream? Kill off employment-generating private enterprise and produce more government jobs. Estimates for the rise in electricity prices are between 4% and 17%, the elderly and those in rural US will suffer even more, jobs will be outsourced abroad, jobs growth will be suppressed, business will be harmed, disposable incomes will be lowered and everyone will feel warm and fuzzy because the US has, yet again, saved the world.

The US system has many checks and balances to disperse power. The political system will be working hard to derail yet another disastrous Obama economic policy. The Democrats are claiming that the Clean Power Plan is not an erosion of constitutional processes and the Republicans need a hymn sheet from which they all sing. The number of disenfranchised voters is high. The Democrat candidate Hillary Clinton is trying to do a balancing act reconciling Obama's plan with her rhetoric of *"fighting for the middle class"*. She can't have it both ways. The new Clean Power Plan regulations were not passed as legislation and there will be a fight in the courts that may spill over into the ballot box. And how will the US generate the required electricity to remain a growing economy? It can only be by gas as it takes many years to build nuclear and hydro plants. It certainly will not be from renewable energy for all the well-known reasons.

The oil and gas boom

Predictions about gas and oil resources don't have a good record. For example, in 1922 a US Presidential Commission[193] stated that "*already the output of gas had begun to wane. Production of oil cannot long maintain its present rate*". It did. President Carter announced in 1977 that "*we could use up all the proven reserves of oil in the entire world by the end of the next decade*". We didn't. The exact opposite happened. There is an oil boom in the US.

Shale oil and gas are undergoing an unheralded productivity boom in the USA.[194] A 321% increase in productivity over four years will keep energy prices depressed and only the lowest cost most efficient producers will be able to be profitable. This is no accident. There is still creativity and a spirit of entrepreneurship in the US. In 2014, I visited a rig in Texas drilling a horizontal hole for shale oil and the technology used then is now out of date. We have yet to see how long-term low petroleum prices and over production will affect the Middle East, Russia, Nigeria and Venezuela but it doesn't look pretty. The current shale oil and gas boom in the US is on the scale of the IT boom and both still have a long way to run.

The boom is due to fracking, which has increased the production of oil and gas from "tight" sequences (i.e. sequences where the porosity and permeability needs to be induced to release trapped hydrocarbons in thin sand layers, from between grains and in tight fractures). Changes in technology such as rig efficiency, geonaviga-

193: http://www.rationaloptimist.com/blog/fossil-fuels-are-not-yet-finished-not-obsolete-nor-a-bad-thing-(1).aspx
194: James Phillips, *City Wire*, 26 August 2015

tion and petrophysics have assisted the fossil fuel revolution. This revolution took place because of free markets, not because of government directives, increased taxation, subsidies or retrogressive energy policies. Although hydrocarbon prices have been falling, drilling and fracking costs have also been falling.

It was only a few years ago that the US was a net importer of liquid hydrocarbon fuels. As a result of the dash for gas, the US is now poised to flood world markets with unthinkable quantities of liquefied natural gas (LNG). The tight oil produced in the fossil fuel revolution has now made the US self-sufficient. It does not need to rely on crude oil from the Middle East. It does not need to play the games of OPEC. The US, a country with the oldest and most developed oil and gas fields in the world, is again at the top of the league rivalling Saudi Arabia and Russia. It is too early to see how this will play out with Middle Eastern and Russian trade, politics and conflagrations. Four LNG terminals under construction will be exporting gas in 2016 and it is probable that the US will overtake Qatar as the world's largest LNG exporter. Australia is also in this race and should overtake Qatar when the offshore Gorgon field comes on stream.

Despite the shale oil and gas revolution that transformed the US from a net importer of oil to an exporter, Obama placed obstacles in the way of new pipelines and stopped the search for shale oil on Federal lands. Although the US economy is recovering at a meagre 2% growth rate, the US GDP is still lower than that of China.

The world is changing rapidly. Detroit was once the home of cars, China now produces twice as many cars as the US and its steel production is nine times that of the US. The low electricity prices that gave the US a great international advantage have now been sabotaged by the green left political policies on climate by Obama.

Down under

Climate change has been a poisonous political policy. In Australia it led to the loss of power by two opposition leaders, two prime ministers, an election loss and political party divisions. Many still get nightmares and cringe about the Australian Prime Minister Kevin Rudd claiming that climate change is the *"great moral, environmental and economic challenge of our age."* This was done just before the Copenhagen climate conference[195] and Obama is doing exactly the same just before the Paris climate conference.[196] Rudd's legacy is that he was exposed as a sham and became the laughing stock of Australia. It was Copenhagen that led to Rudd's removal in his first term as Prime Minster.

What if I am wrong and a reduction of CO_2 emissions is absolutely necessary to save the planet? This is a question that most scientists ask themselves. If Australia stopped all CO_2 emissions today, global temperature would decrease by 0.0154°C by 2050. Such a wonderful planet saving exercise has absolutely no real effect on global temperature (even if one assumes incorrectly that CO_2 drives global warming). Not only would Australia become bankrupt but it could not feed itself. Such voluntary acts of international environmental kindness would have absolutely no effect on the global climate. Green left environmental activists may claim that we should lead by example. Maybe the Green Party would willingly commit economic suicide; no one else would least of all developing countries like China and India. They want our standard of living and nothing is going to stop them.

The Copenhagen Accord was drafted by the US, China, India, Brazil and South Africa on 18 December 2009, it was "taken note of", was not passed unanimously, the document was not legally bind-

195: 2009 UN Climate Change Conference, Copenhagen 7-18 December 2009
196: COP21, Paris, 30 November -11 December 2015

ing and did not contain any legally binding commitments to reduce human emissions of CO_2. So why bother? The Copenhagen talk-fest, at which Prime Minister Rudd and more than a hundred Australian bureaucrats attended, achieved absolutely nothing and cost a king's ransom. Who cares? It's only taxpayers' money. Rudd thought he could walk away as an international hero who had saved the world from themselves. He was humiliated. Obama and the Pope are heading down the same path with Paris.

In response to my 2011 book,[197] the Gillard Labor Government's oxymoronic Department of Climate Change and Energy Efficiency established a web site[198] (now removed after a 7 September 2013 change from a Labor to a Conservative government) entitled *"Accurate Answers to Professor Plimer's 101 Climate Change Science Questions."* They claimed the web site was *"based on up-to-date peer reviewed science, and have been reviewed by a number of Australian climate scientists"*. I didn't know whether to laugh or cry because the document read like one of the many anti-science complaining creationist critiques that I have read (under sufferance). It was clear that my book had stung the catastrophic climate clique, was helping change community opinion and the web site was an act of desperation. It showed. One correspondent calculated that to set up such a site would have cost the taxpayer $1.5 million.

It was amusing that there were 101 questions for pupils to ask their teachers and even this group of ever-so-eminent and serious folk was unable to answer many of the questions posed. This was the politicised science at its worst. We can't just have a University Chair publishing ideas contrary to government policy, can we? If we can't sack him, then let's set up a site at taxpayer expense to demonise

---

197: Plimer, Ian 2011: *How to get expelled from school: A guide to climate change for pupils, parents and punters.* Connor Court
198: http://www.climatechange.gov.au/climate-change/understanding-climate-change/response-to-prof-plimer.aspx

him. The web site was loved by the normal primitive suspects in the media, uncritical environmental activist groups and the lunar left. And the taxpayer paid for this blatant green left political propaganda. There was nothing the Labor government could do to me because I did not depend upon government research grants to continue my science. All the Labor government could do was to, in effect, publicly brand me an enemy of the state. This is a tactic used in totalitarian countries.

It was the same Obama who lectured Australia at the time of the G20 meetings in Brisbane about how the Great Barrier Reef has been destroyed because of CO_2 emissions by humans. Now it has been proved that the Great Barrier Reef was never threatened, we just see how hollow the words of Obama were. And how did Obama get to Australia? In the Boeing 747 called Airforce One that flies by burning kerosene and emitting CO_2. At any one time, about half a million people are in the air in about 14,000 aircraft with a Boeing 747 burning about 175 tonnes of kerosene between refuels. An A380 Airbus burns 310 tonnes between refuels. Commercial airliners average 2.3 cycles per 24-hour day. About a trillion tonnes of kerosene are burnt to CO_2 each year by aeroplanes. How do thousands of activists attend a Rio, Copenhagen or Paris talk-fest? By international air travel.

We mere mortals are sinning when we fly whereas the hand wringers trying to save us from ourselves can justify their aeroplane CO_2 emissions because they are on a pilgrimage. If the green left environmental activists were not hypocrites, then they would have attended these international talk fests by using renewable energy rather than the generous use of CO_2-emitting aeroplanes. The Paris meeting is in late November-early December 2015. Have any German green left activists now started to retrace Napoleon's steps and walk to Paris for the UN meeting? And how does the Pope travel to meet his flock in South America, Philippines or Africa? By aeroplane.

Old Dart

The UK was once the global coal mining centre. Almost all UK coal mines are closed. The UK generates electricity from coal, nuclear, biomass and renewables. The UK is facing an energy and revenue crisis. Half a century after the establishment of the North Sea oil industry, the main oil fields are exhausted. The remaining small resources are probably not worth the investment by the big players. Three years ago, North Sea oil provided £11.5 billion to the UK government, now it provides £1 billion. Between 2020 and 2040, the total royalty will be £2 billion, which amounts to a 94% reduction on previous forecasts. The North Sea oil will shift from being an asset to a liability because the UK taxpayer will have to pay 60% of the £30 billion decommissioning costs.[199] The energy policy in the UK is a mess.

In the UK, if fracking took place in Lancashire and Yorkshire, it is estimated that the UK could generate all of its electricity for the next 43 years. This would generate 60,000 jobs and billions of pounds of investment.[200] The UK has now decided to fast track fracking. When the fracking industry becomes global, the world will change dramatically. Texas and many of the US states with hydrocarbon basins have changed very quickly. Maybe the same will occur in the UK.

The developing world

In 2013, China accounted for 47% of global coal consumption.[201] China emits more than 400 million tonnes of CO_2 each month that is about 45% more than Obama's Clean Power Plan would "save" in a year. However, the Chinese emissions of CO_2 may have been over estimated by at least 14%. Just to make it a little more interesting, China under-reported its coal consumption by up to 10%. Despite this, China's emissions between 2000 and 2013 were almost three bil-

---

199: *The Sunday Times*, 23 August 2015
200: *The Times*, 10 August 2015
201: US Energy Information Administration

lion tonnes less than official estimates. This means that when China burns coal, it produces about 40% less CO_2 than models assume.[202] Furthermore, Chinese cement production has been over-estimated by up to 45%. Who really knows what goes on in China? What we do know is that these incorrect estimates are based on default calculations used by the IPCC.

India's coal burning is also increasing and will continue to increase. There was a time when US action drove other countries into taking similar action. However, in what is now an industrial revolution in East Asia and the Indian sub-continent, nothing the US does can *"save the planet"* because China and India are not going to sacrifice economic growth for an unfounded ideology. QED.[203] Why even spend millions on a Paris conference?

Coal consumption in India, particularly in the electric power sector, is outpacing India's domestic production. No wonder Indian coal-fired electricity generating companies such as Adani and GVK are looking to Australia to supply the coal. India's coal production grew only 4.7% between 2005 and 2012 (600 million tonnes) whereas power capacity grew at 9.4% per annum reaching 150 GW.[204] India has set a coal production target of 1.5 billion tonnes by 2020. Despite pressure before Paris, India refuses to indicate to the UN when its CO_2 emissions will peak.[205]

The Parisian junket

Thousands of concerned folk will fly into Paris. Their carbon footprint will be ignored because they are working for the greater good of mankind. For you, for me and for our present or future children

---

202: Guan, D. *et al.* 2011: A gigatonne gap in China's carbon dioxide inventories. *Nature Climate Change* 2: 672-675

203: *Quod erat demonstrandum*

204: US Energy Administration, 25 August 2015

205: *The Times of India*, 25 August 2015

and grandchildren. Every four years we fund these folk to add CO_2 to the atmosphere to travel to some far-flung very seriously important conference venue. There will be Parisian huffing and puffing, all night negotiations and no binding meaningful agreement despite the attempt to produce a global bureaucracy to limit CO_2 emissions. On these junkets, delegates of course would not touch an alcoholic drink because CO_2 is released during fermentation. No soft drink would be consumed because they contain CO_2. It's a hard life being an unappointed moral guardian.

Politicians and bureaucrats look for feel good quasi-successful fudges and then can report success. To produce a failure but call it a success is one of the oldest political tricks in the book and this is the expected best-case outcome for Paris. There may even be a legally binding agreement to disagree about CO_2 emissions whereby each country is legally bound to do whatever it wants and, if they wish, to set their own emissions targets. This is what happened at Copenhagen. This is the escape from the climate treaty dilemma.

A few potholes have appeared in the road to Paris. The green left environmental activists are probably not aware that there has been a change in the government in the UK. The Greens, Labour and socialists lost heavily, the Conservative government is stopping subsidies for wind power, there is a focus on the high consumer cost of electricity and the Department of Energy and Climate Change may be dismantled.[206]

The US Congress will disallow any legally binding international agreement[207] and the Australian Government has directed that the government-funded Clean Energy Finance Corporation stop funding wind power or small scale solar installations.[208]European finances

206: http://www.telegraph.co.uk/finance/budget/11718594/Green-energy-subsidies-spiral-out-of-control.html
207: http://www.theguardian.com/world/2015/jun/01/un-climate-talks-deal-us-congress
208: http://smh.com.au/federal-politics/political-news/government-pulls-the-plug-on-household-solar-20150712-gianOu.html

are not too healthy (especially in Greece),[209] 92% of Swiss voters rejected a carbon tax in March 2015. The German government dropped its proposed carbon tax on coal-fired power stations and was forced to guarantee backup for wind and solar when they didn't perform.[210] The developing world, especially China and India, will agree to nothing that curbs their growth and will make hollow pledges for some trivial meaningless action in decades time.

The developing nations are not stupid. Paris not about climate, it's about easy money. Guilt money. They have tried to set a trap for Western green politicians who have set the trap for themselves. They are demanding $200 to $400 billion dollars *per annum* for so-called climate compensation and adaption measures. The Pacific Islands Development Forum (PIDF) will call on the world's biggest CO_2 emitters to compensate all Pacific Islands that are affected by climate change. PIDF Interim Secretary-General Amena Yauvoli said:

Those responsible for emitting the most greenhouse gases should pay.

If Paris succeeds against all odds with the green left environmental activists' dream and produces an internationally binding climate treaty, then a new unelected bureaucracy will be able to interfere with any country and override the national government. We already have this in the EU where bureaucrats in Brussels can overrule elected members of parliament in Westminster. No wonder the UK Independence Party (UKIP) appeared and grew so rapidly. If Paris achieves what the green left environmental activists desire, there will be a great loss of national sovereignty, far greater than any invasion or war. And all we hear in the media is promotion of incorrect predictions about increased warming, increased extreme weather, the disappearance of polar ice, extinction and threats to polar bears.

209: http://www.energytribune.com/10083/greek-debt-crisis-exposes-green-energy-subsidies#sthash.JMBqVEQW.dpbs
210: http://blogs.ft.com/nick-butler/2015/07/06/germany-the-coal-industry-lives-on/

When Al Gore was born, there were 7,000 polar bears on Earth. Today, only 26,000 remain.

Do we ever hear the media expressing concerns about democracy and sovereignty when writing about the forthcoming UN meeting in Paris?

3

CARBON DIOXIDE, SCIENCE AND CLIMATE

THE FOOD OF LIFE

On the basis that CO_2 emissions drive climate change, a study at Concordia University (Montreal, Canada)[211] shows the ranking of the top 20 contributors computed from fossil fuel emissions, land use CO_2 and aerosols.

The proportion of annual global emissions is USA (15.1%), China (6.3%), Russia (5.9%), Brazil (4.9%), India (4.7%), Germany (3.3%), UK (3.2%), France (1.6%), Indonesia (1.5%), Canada (1.4%), Japan (1.3%), Mexico (1.0%), Thailand (0.9%), Columbia (0.9%), Argentina (0.9%), Poland (0.7%), Nigeria (0.7%), Venezuela (0.7%), Australia (0.6%) and The Netherlands (0.6%).

The paper also shows that there are vast disparities in both total and *per capita* "climate contribution" among countries and, that across most developed countries, *per capita* contributions are not currently consistent with attempts to restrict global temperature change to less than 2°C above pre-industrial temperatures. Why 2°C? How was this figure derived? Why not 1°C or 3°C? How can we humans twiddle the planetary dials to adjust the temperature?

The bottom line

Carbon dioxide feeds plants and plants feed us. Whether we are vegan, vegetarian, omnivorous or carnivorous we need plants to stay alive. Breatharians don't, or so they tell us. Without CO_2 or even a

---

211: Matthews, D. H. *et al.* 2010: National contributions to observed global warming. *Envir. Res. Lett.* 9: doi:10.1088/1748-9326/9/1/014010

significantly reduced atmospheric CO_2 content, there would be no life on Earth.

To argue for a reduction of human emissions of CO_2 or sequestration of CO_2 at the site of a coal-fired power station is proof that the education system has failed. Put up your hands. Who has heard of photosynthesis?

Demonisation of plant food

That colourless odourless tasteless non-poisonous gas called CO_2 is lurking in the atmosphere. It is so terribly dangerous that we have decided to bury it by using wonderfully sounding projects called "carbon sequestration". Never mind that when plants and animals grow they also sequester carbon. We have known since 1804 AD that CO_2 was plant food[212] and the father of the modern global warming theory Svante Arrhenius argued in a paper that he gave to the Stockholm Physical Society called "The influence of carbonic acid in the air upon temperature of the ground" that with increasing fossil fuel use, there would be an increase in the rate of plant growth.[213]

Arrhenius argued that a higher CO_2 content in atmosphere would be of benefit to man and had no concern about a potential slight warming. And why would someone in cold Sweden be worried about a slightly better climate? Today the green left environmental activists argue the opposite and tell us that agriculture, ecosystems and life will be destroyed by a slight increase in atmospheric CO_2 despite no change in CO_2 chemistry during the last 100 years.

Although CO_2 is a trace gas in the atmosphere and is plant food, we have to bury the evil CO_2 we produce as a result of keeping humans employed, warm, cool, fed and watered. Sequestration aims to keep the human emissions of CO_2 away from the natural emissions

---

212: Nicolas-Théodore de Saussure, 1804: *Recherches Chimiques sur la Végétation*. Nyon
213: Svante Arrhenius, 1908: *Worlds in the making: the evolution of the universe*. Harper

of CO_2 in the atmosphere. Why? From the green perspective, a very slight increase of CO_2 in the atmosphere can have great effects on climate and weather but only if it is from sinful human emissions and not from natural emissions.

It seems this miracle gas can simultaneously cause extreme heat and extreme cold, flooding rain and endless drought, increased snow and a lack of snow, increased wind and a lack of wind, and increased hurricanes and a lack of hurricanes. Or so we are told. This is not a bad score sheet for a trace gas. We are bombarded with the media propaganda that whatever change is observed on Earth, it is due to human emissions of CO_2. The real miracle of CO_2 is that, without this trace gas, there would be no life on Earth. This seems to have escaped the attention of green left environmental activists. Maybe it hasn't. Maybe they are being disingenuous with the scientifically illiterate media.

In the Australian Parliament, the Minister for Agriculture, Barnaby Joyce claimed that the expropriation of farms for carbon sequestration had cost farmers $200 billion in land values.[214] This figure was not challenged by the Green Party and their fellow travelers on the left.

To legislate for coal-fired power stations to bury the CO_2 they produce makes coal-fired electricity generation so inordinately expensive that subsidised wind- and solar-generated electricity becomes relatively competitive. This is why renewable proponents can take the high moral ground and gloat that wind-powered electricity is cheaper than coal-fired electricity.

The wonder gas

Carbon dioxide is plant food. It is used to produce the organic matter out of which plants construct tissues. Horticulturalists have known for a long time that pumping warm CO_2 into glasshouses increases

---

214: http://catallaxyfiles.com/2015/08/08/land-theft-by-governments-the-judiciarys-mythical-defence-of-darryl-kerrigans-castle/

yields. It appears that in a glass house a CO_2 content of 1,600 ppm is beneficial[215] yet we are told by green left environmental activists that increased CO_2 in the atmosphere to 400 ppm is dangerous. Furthermore, the amount of water used by plants decreases with increased CO_2 because stomata are open for a shorter time hence water transpiration is reduced. Do they realise that when they are indoors in winter there can be up to 1,000 ppm CO_2 in the air, which is why pot plants thrive indoors.

Addition of CO_2 to the atmosphere can feed more people. Of 45 crops that account for 95% of global crop population, the Center for the Study of Carbon Dioxide and Global Change has shown that an increase of 300 ppm CO_2 increases yields by 5 to 78%, improve the efficiency of plant water use and increases a plant's ability to withstand drought.[216] Similar increase in yields have been shown from other studies. The US estimated gains in yields from 1950 to 2009 due to CO_2 fertilisation were cotton (51%), soybeans (15%), wheat (17%), corn (9%) and sorghum (1%).[217] The same is seen elsewhere for rice.[218] Most crops and plants are C3 plants, probably because they evolved at a time when atmospheric CO_2 was far higher than now. Many weeds are C4 plants and higher atmospheric CO_2 inhibits weed growth.[219]

Green left environmental activists tell us that CO_2 is dangerous

215: Mortensen, L. M. 1987: Review: CO_2 enrichment in greenhouses: crop responses. *Scientia Horticulturae* 33: 1-25

216: Idso, C. 2013: The positive externalities of carbon dioxide: Estimating the monetary benefits of rising atmospheric CO_2 concentrations on global food production, http://web.uvic.ca/~kooten/Agriculture/CO2FoodBenefit(2013).pdf

217: Attavanich, W. and McCarl, B. A. 2014: How is CO_2 affecting yields and technological progress? A statistical analysis. *Climate Change* doi 10.1007/s10584-014-1128-x

218: von Caemmerer, S. *et al.* 2012: The development of C4 rice: Current progress and future challenges. *Science* 336: 1671-1672

219: Zeng, Q. *et al.* 2011: Elevated CO_2 effects on nutrient competition between a C3 crop (Oryza sativa L.) and a C4 weed (Echinochloa crusgalli L.). *Nutrient Cycling in Agroecosystems* 89: 93-104

and will lead to runaway global warming. They carp on about "carbon" pollution. This is calculated deceit. Carbon is black and solid. When atmospheric CO_2 was up to 1,000 times higher than now in the past, there was no runaway global warming. Life did not die. Plants thrived. In the US, the EPA has even classified CO_2 as a pollutant. This is misleading and deceptive. One legal stroke by activists in the EPA has erased the whole history of life on the planet. All life is carbon based and without CO_2 there is no life on Earth. All carbon in oil, coal and gas that we burn to produce CO_2 was once in the atmosphere. Fossil fuels ultimately result from solar energy.

To feed a population of maybe 8 or 10 billion people, we need to keep the wheels of the green revolution spinning. However, all crops are engaged in a perpetual growing season war of attrition with fungal parasites, insect predators and plant competitors (weeds). These wars are assisted by more atmospheric CO_2, better GM crops, better fertilisers, insecticides and herbicides and cheap electricity. As Western countries became more efficient, fewer people lived in rural areas, forests expanded, food productivity became more efficient, the birth rate declined, longevity increased and employment based on cheap energy became concentrated in the cities.

Simulations of natural sequestration of CO_2 for the period 1960 to 2009[220] in the tundra and forests of northern Eurasia show that emissions of CO_2 are overestimated whereas uptake of CO_2 by plants is underestimated. More CO_2 is stored than emitted by human activities. Simulations suffer from the lack of ground measuring stations in remote areas and hence, as with so many other climate studies, there is uncertainty. This is in accord with the IPCC's assumption that has minimised by a factor of 10 the way in which CO_2 is removed from the air.[221]

220: Rawlins, M. A. *et al.* 2015: Assessment of model assessments of land-atmosphere CO_2 exchange across Northern Eurasia. *Biogeosciences* 12: 4385-4405
221: http://hockeyschtick.blogspot.com.au/2013/05/analysis-finds-co2-emissions-only.html

The IPCC Bern model calculates that 22% of all atmospheric CO_2 surplus remains in the air. If carbon is artificially produced, such as the C^{14} of an atomic bomb, it can easily be measured that more than 95% of the C^{14} has already been removed after 50 years. The life of a molecule of CO_2 in the atmosphere is short. If human emissions of CO_2 don't remain in the atmosphere for very long and are naturally sequestered into life, then what's the problem? The planet has been naturally sequestering CO_2 for thousands of millions of years and just because we humans are on Earth does not mean that major planetary recycling systems change.

The Earth is getting greener.[222] Satellite measurements have confirmed that green vegetation on the planet has been increasing for three decades. Furthermore, the increased greening detected via satellite and aircraft measurements is consistent with the increase in global crop yields for more than 50 years.[223,224] The world's forests have increased in area[225] across all climate zones[226] and are growing faster than they did 50 years ago.[227] Had it not been for an increase in crop yields of 9 to 15%, then global croplands would have had to be increased in order to produce the same amount of food. These measurements are contrary to the Pope's opinion about the alarming rate of deforestation, over-development and ecosystem destruction.

222: Liu, Y. Y et al. 2015: Recent reversal in loss of global terrestrial biomass. *Nature Climate Change* 5: 470-474

223: Food and Agriculture Organization of the United Nations, http://www.fasostat.fao.org

224: Zeng, N. et al. 2014: Agricultural green revolution as a driver of increasing atmospheric CO_2 seasonal amplitude. *Nature* 515: 394-397

225: Pan, Y. et al. 2011: A large and persistent carbon sink in the world's forests. *Science* 333: 988-993

226: Donohue, R. J. et al. 2013: CO_2 fertilization has increased maximum foliage cover across the globe's warm, arid environments. *Geophys Res Letts* doi 10.1002/grl.50563

227: Pretzsch, H. et al. 2014: Forest stand growth dynamics in Central Europe have accelerated since 1870. *Nature Communications* 5: doi 10.1038/ncomms5967

This greening has added nearly four billion tonnes of carbon to land plants and an unknown amount to microorganisms since 2003. Tree planting in China, forest regrowth in the former Soviet Union, less land clearing from better agricultural practices and a slightly higher atmospheric CO_2 content are making the planet a better place.

Satellite pictures of North Korea show massive deforestation and no greening due to CO_2. Night-time pictures show the lack of electricity in North Korea. This is the modern face of poverty and totalitarianism.[228] By contrast, we see the overall greening of planet Earth from satellites.[229]

Why would the United Nations want to reduce human emissions of CO_2 that have demonstrably raised the standard of living for so many of the world's poor? It seems the focus is an attack on Western industrialisation rather than an attempt to stop poverty.

CLIMATE PREDICTIONS

Green left environmental activists and their uncritical supporting media[230] are engaged in emotional posturing, selective use of information, omission of critical data, moralising, crying wolf and predicting doom and gloom. They have predicted 60 of the last 10 crises. Global average sea surface temperatures have not increased in any significant way since 1998,[231] average global temperature has not increased for more than 18 years,[232] Arctic sea ice is increasing[233]

228: www.droyspencer.com
229: Zhou, L.M. *et al.* 2001. Variations in northern vegetation activity inferred from satellite data of vegetation index during 1981 to 1999. *Jour. Geophys. Res.*106: 20,069-20,083
230: Kenny, Chris, 2015: Media Watch's climate change obsession. *The Australian*, 1-2 August
231: http://judithcurry.com
232: http:// www.forbes.com/sites/jamesconca/2015/06/15/a-pause-in-global-warming-not-really/
233: http://nsidc.org/arcticseaicenews/

and Antarctic sea ice is increasing.[234] The Arctic sea ice area initially decreased and now it is back to the 2006 levels.[235] However, green left environmental activists see it differently and try to present heroic arguments that all is not well and we are doomed.

Their scary predictions come from models, not measurements. I think that there is a significant difference. Modellers actually don't collect primary data themselves and their claim to fame is to massage the data collected by others. Matthew England[236] of the Climate Change Research Centre (UNSW) is one of these and continues to amuse us:

> *What we're seeing in models is that the warming out of the hiatus is gonna be rapid, regardless of when the hiatus ends.*

Forget the observations, have blind unreasoning faith in the models. Considering that the models can't be run backwards to show what we have measured, that the models did not predict the hiatus and it now appears that those sneaky models can predict when the temperature hiatus will end. And you better believe it folks, it's going to be fast, furious and final. One can only believe this stuff when in a strong embrace with Bacchus. If that's all that we get from a taxpayer-funded climate institute, then it should be shut down.

History of CO_2

Limestone and limey rocks contain calcite ($CaCO_3$), a mineral that contains 44% CO_2 by weight. Burial of carbon compounds prevents conversion of carbon compounds to CO_2 and methane that could be released to the atmosphere. Burial of carbon compounds was accelerated at about 400 Ma when land plants first appeared on the

234: https://www.nasa.gov/content/goddard/antarctic-sea-ice-reaches-new-record-maximum/
235: https://www.arctic.moaa.gov/detect/ice-seaice.shtml
236: Kenny, Chris, 2015: Media Watch's climate change obsession. *The Australian*, 1-2 August, p. 19

planet. Forests grew quickly, CO_2 was removed from the atmosphere and carbon was not recycled as atmospheric CO_2 because it was buried as coals, carbonaceous sediments, limey sediments and limestone reefs.

There were times, such as the Carboniferous, when there was an explosion of plant life on Earth. There was a massive removal of CO_2 from the atmosphere, further oxygenation and storage of recycled carbon in Northern Hemisphere coals. The removal of CO_2 from the atmosphere was immediately before the Permo-Carboniferous glaciation and may well have been one of the factors that set the stage for an ice age. At that time, the supercontinent Gondwana was at the South Pole and the Permo-Carboniferous ice sheets were at high latitudes.

If the Permo-Carboniferous glaciation was influenced by a low atmospheric CO_2 content, then what about the glaciation which we now enjoy? During the break-up of the supercontinent Gondwana over the last 100 million years, India drifted northwards and collided with Asia 50 million years ago. Australia got rid of New Zealand 100 million years ago. And a good thing too. The Himalayas were pushed up. India is still pushing against Asia and the Himalayas are still rising. Every time there is an earthquake in Tibet, Nepal, northeastern Pakistan, northern India and southwestern China, it is because the Himalayas are still rising and break rocks to do so.

Accumulation of high-altitude snow and ice increased the amount of reflected solar energy. The area of the Himalayas is about half that of the USA so the feedback effect of reflected solar energy is large. The size and height of the Tibetan Plateau changed global wind patterns, resulting in regional changes to climate. The great difference between winter and summer temperatures shatters rocks. This produces pieces of rock with a large surface area for attack by rain and microorganisms. The huge Tibetan Plateau heats in summer, the air above it heats and rises and cooler moist air from the tropical Indian Ocean is dragged up to the plateau. This results in monsoonal rains that attack the rocks to form soils, a process that removes H_2O, oxygen and CO_2 from the atmosphere and adds salts and bicarbonate to the oceans.

Soils are stripped from the steep slopes in periods of high rainfall and the process starts all over again. The thick pile of sediments at the Ganges Delta in the Bay of Bengal shows that this process of CO_2 removal from the atmosphere has been taking place for 50 million years. Some 15 million years after this process started, ice appeared on Antarctica.[237]

There is a hypothesis that the uplift of mountains changes the rate of weathering by removal of CO_2 from the atmosphere to produce a negative greenhouse effect. The theory is that if an increase in atmospheric CO_2 drives global warming, then a decrease in atmospheric CO_2 should drive cooling. However, there is no correlation between mountain building and glaciation hence there is no relationship between atmospheric CO_2 and global temperature. But we knew this anyway.

Increased mountain building increases the draw down of CO_2. It has been proposed that the uplift of the Himalayas was one of the climate drivers that triggered the latest glaciation.[238,239] This conclusion is hotly contested. Increased weathering adds more nutrients to the oceans. For example, the rapid uplift of the Himalayan-Tibetan Plateau led to an increased input of phosphorus to the oceans. This led to a blooming of algae in the oceans between eight and four million years ago coincidental with the intensification of the Indian-Asian monsoon. Not only did the uplift of mountains draw down

237: Raymo, M. E. and Ruddiman, W. F. 1992: Tectonic forcing of late Cenozoic climate. *Nature* 359: 117-122

238: Ruddiman, W. F. and Kutzbach, J. E. 1991: Plateau uplift and climatic change. *Scientific American*, March 1991: 42-50

239: Raymo, M. E. and Ruddiman, W. F. 1992: Tectonic forcing of late Cenozoic climate. *Nature* 359: 117-122

CO_2 from the atmosphere but algal blooming[240] also further depleted atmospheric CO_2.

The cooling Sun

The Sun drives all processes at the surface of the Earth, including weather and climate. Sunlight drives photosynthesis. It is photosynthesis that gave us coal, oil, gas and modern plant life. Without the Sun, there is no life on Earth (although a few extremophiles may lurk around volcanic hot springs). Solar energy heats the oceans and ocean temperature drives the atmospheric temperature. Slightly warmer oceans release more CO_2 to the atmosphere than cooler oceans.

The Sun contains 99.86% of the mass in the Solar System. The Sun's output is variable. Not since Solar Cycle 14 peak in February 1906 has there been a solar cycle with fewer sunspots.[241] Going back to 1755 AD, there have only been a few solar cycles that have had a lower number of sunspots during their maximum phase. This continued downward trend in solar sunspot cycles began at least 18 years ago when the Earth stopped warming. If it continues for a couple more cycles, the Earth could be entering a Grand Minimum with an extended period of low solar activity. Many are now arguing that the lack of warming since 1998 is caused by a lack of solar activity[242] and that weakening solar activity could bring on another Little Ice Age.[243]

From the geological perspective, the next cooling is inevitable because of the Earth's orbital perturbations and variable solar activity.

240: Excess surface water nutrients or eutrophication, commonly by marine algae called coccolithophoridae.
241: http://dailycaller.com/2015/04/30/the-sun-is-blank-as-solar-activity-comes-to-a-standstill/
242: http://www.eike-klima-energie.eu/climategate-anzeige/waermebilanz-der-erde-und-globale-temperaturaenderung/
243: http://dailycaller.com/2015/03/03/paper-global-warming-more-like-global-cooling/

However, it is not known whether this is decades, centuries or millennia away. It certainly will be on a Thursday.

Some caution needs to be exercised because, in the 1970s, the same scientists who now tell us that we will fry and die were telling us that the Earth is heading for a mini ice age.[244] *National Geographic*,[245,246] *Time*[247] and *Newsweek*[248] had front page lead stories telling us that we were heading into a new ice age because atmospheric temperature had been decreasing. They didn't tell us that atmospheric CO_2 was increasing. In 2015, *Newsweek* had an article written by an atmospheric scientist telling the world that a solar-driven mini ice age is bogus. *National Geographic*,[249] *Time*[250] and *Newsweek*[251] have had many headlines telling us that we'll fry and die. Now the popular media tells us that we are heading for catastrophic global warming because atmospheric CO_2 is increasing. In 1975 we were told:[252]

> *Global cooling presents humankind with the most important social, political and adaptive challenge we have had to deal with for 110,000 years. Your stake in the decisions we make concerning it is of ultimate importance: the survival of ourselves, our children, our species.*

All pretty scary stuff. Just substitute the word warming for cooling and this statement could have been made by any of today's green left environmental activists. It is clear that the popular media do not understand science and just want to sell scary stories. When reading

244: Lowell Ponte, 1976: *The Cooling*. Prentice Hall
245: *National Geographic*, November 1976
246: *National Geographic*, October 2010
247: *Time*, April 1977
248: *Newsweek*, 28 April 1975
249: *National Geographic*, September 2004
250: *Time*, April 2006, April 2007, April 2008
251: *Newsweek*, 22 January 1996
252: *Newsweek*, 28 April 1975

the popular press, keep in mind the quote from the Duke of Wellington, "*If you believe that you will believe anything.*"

Planet Earth warmed the end of the last glaciation about 12,000 years ago through to a peak some 6,000 years ago known as the Holocene Optimum. Within this 12,000-year long warm period there were some coolings (e.g. Younger Dryas). Since the 6,000 year peak some 2°C warmer than now, we have seen gradual cooling but punctuated by a cyclical pattern of warmer periods like the Roman Warming and cooler periods like the Little Ice Age. The zenith of the Little Ice Age was 350 years ago, since then there has been a long-term trend of warming within which there have been cooling and warming events as well as periods of no change, all of which are unrelated to human emissions of CO_2.

Maybe climate predictions have nothing to do with climate or science. Because billions of dollars is spent each day in the forlorn hope of trying to change the weather, there might be just a little self-interest involved and we might not be getting data collected, validated, repeated and interpreted dispassionately. History will not judge us kindly. During a time when 20% of the world has no electricity, we have the echo chamber of politicised and bureaucratised science such that politicians, the crowded green left space of environmental journalism and socialists are getting the answers that suit their ideology. The end result is that the cost of energy has risen, the poor are having increasing difficulty in affording energy and the process of science has been compromised.

The elephant in the room is: What happens if we are wrong? This is a question I have often asked myself in my scientific career. And I have been wrong. And I have published papers criticising my earlier work. During the ideological maelstrom of exaggeration, Climategate, adjusting primary data, fraud, selectivity of data and demonising dissent to prove human-induced global warming, who is asking the simple questions? What happens if we are wrong? What happens if the surface of the planet is cooling and not warming? Just because human activity emits larger CO_2, why shouldn't the Earth enter its next inevitable orbitally- and solar-driven cool period? This would, as in the past, result in desertification, advance of ice sheets, retreat of forests, a reduction of arable land, species loss and reduction in species numbers.

Despite Western government-funded scientists, scientific societies, media networks and some politicians claiming that the Earth's climate is continuing to get warmer, there is just a little problem with data. The attempts to change data have not changed the facts. There has been no warming of the atmosphere for over 18 years and whatever the explanations offered, the facts remain the same.

Questions are being asked. Are the Northern Hemisphere winters getting colder? Are cold weather extremes in the Northern Hemisphere such as increasing snow accumulation and extreme cold spells over the last five years evidence of cooling? India and NE Japan have experienced colder and snowier winters compared to the 1990s. Winters are becoming colder over certain parts of the land in the Northern Hemisphere. Cooling may be about 0.6°C since 2000 and temperature trends in many parts of the world are downward.[253,254] It is clear why some organisations feel the need to adjust these downward trends that don't fit the ideology.

It is 172 years since it was realised that the Sun's activity varied over a cycle of around 10 to 12 years.[255] Every cycle is slightly different. It was initially thought that the cause was a magnetic dynamo caused by a deep convection fluid. A new and better interpretation is that cycles can be explained by using a dual dynamo.[256] The Sun is

253: http://icecap.us/index.php/go/joes-blog
254: NOAA average temperature chart; http://www.ncdc.noaa.gov/cag/
255: In 1843 after 17 years of observations by Samuel Heinrich Schwabe (1789-1875).
256: Zharkova *et al.* 2015: Irregular heartbeat of the Sun by double dynamo National Astronomy Meeting 2015 of the Royal Astronomical Society, July 2015, Llandudno, Wales

like many other stars. It is a giant nuclear fusion reactor that generates gamma rays, X-rays, ultraviolet light, visible light, infra red rays, microwave radiation and radio waves. It has a powerful magnetic field similar to a dynamo and generates solar particles that bombard Earth. The total solar output from the Sun has only varied by about 0.15% over the last 400 years (i.e. the solar constant); the Sun has essentially a constant output of heat but with variable magnetic fields and variable sunspots.

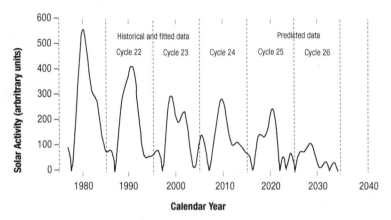

Figure 2: Dual solar dynamos after Zharkova *et al.* 2015

Magnetic wave components of the Sun are a dual-dynamo system that originates from two different layers in the Sun's interior. There also has been a suggestion that there are three solar dynamos,[257] which does not change the basic conclusion: there is a close relationship between solar cycles and climate on Earth and this has been known for decades.[258,259] We may speculate whether the Sun's variable

257: Scafetta, N. 2012: Multi-scale harmonic model for solar and climate cyclical variation throughout the Holocene based on Jupiter-Saturn tidal frequencies plus the 11-year solar dynamo cycle. *Jour. Atmos. Solar Terr. Physics* doi:10.1016/j.jastp.2012.02.016

258: Friis-Christensen, E., and Lassen, K. 1991: Length of the solar cycle: An indicator of solar activity closely associated with climate. *Science* 254: 698-700

259: http://wattsupwiththat.com/2010/02/02solar-cycle-24-update/

UV output or magnetic influence on the Earth's upper atmosphere affects cloud cover thereby changing the amount of energy the Earth reflects back into space. If the Sun affects cloud cover, then the Earth's surface temperature will change despite the Sun emitting almost constant energy. The science is certainly not settled.

It is becoming increasingly popular to predict a forthcoming mini ice age. The Earth has been warming since the Maunder Minimum 350 years ago, this is unrelated to human emissions and is related to changes in solar activity. I know, I know, it is really odd to suggest that the great ball of heat in the sky might have something to do with heat on the surface of the Earth. And it is really scary that the Sun is a nuclear fusion reactor. Is there some legislation that the green left might invoke to stop capitalists gaining benefit from nuclear fusion? To be consistent, the green left environmental activists must demonstrate against the Sun. Carbon dioxide did not begin to start increasing until the 1800s. Whether this is due to human emissions is debatable. The rate of increase grew with the post-World War II industrialisation and the growth of China, India and SE Asia. Hence, any warming before 1800 cannot be due to human emissions of CO_2. This is why the Medieval Warming must be removed from history.

It is quite plausible that the surface of Earth will cool. On one scale, we are in an ice age wherein we are enjoying a 10,000-12,000 year long orbitally-driven interglacial that is drawing to an end. We have seen an overall decrease in temperature since the Holocene Optimum 6,000 years ago. After the interglacial, we can expect a 90,000-year glacial period, unless we have an urgent multi-trillion dollar international agreement to change the Earth's orbit. Don't wait up.

There is a considerable amount of scientific data to show that sunspots can be used to accurately predict the surface temperature on Earth with a lag of one sunspot cycle (~11 years; i.e. half the Sun's cycle).[260,261,262] During peak times the Sun has many sunspots and spits out solar flares. These have previously affected satellites and high voltage power lines. However, the 11-year cycle cannot be used to accurately predict the Sun's behaviour, which at times can appear erratic. This evidence is ignored by green left environmental activists.

If the Earth cools, then we are totally unprepared. Trillions of dollars has been spent on trying to determine whether there is human-induced global warming and how to mitigate the effects of warming. We have been blinded by the shrill propaganda and have not considered the possibility of a cooling Earth. Past warmings and coolings have been due to a coincidence of a number of natural events or cycles.

Scientists from the University of Southampton predict that the Atlantic Ocean is cooling and could cool by as much as by 0.5°C over a few decades. A cooler Atlantic would bring cooler drier summers to the British Isles and drought in the Sahel for a period of 20 to 30 years.[263]

Others are predicting a cooler period based on solar cycles, which could bring another Little Ice Age[264] and argue that the lack of atmospheric temperature rise since 1998 could be due to solar activity.[265] We are now more than six years into Solar Cycle 24 and the low

---

260: Balachandran, N. K. *et al.* 1999: Effects of solar cycle variability on the lower stratosphere and troposphere. *Jour. Geophys. Res.* 104: 27321-26339

261: Scherer, K. *et al.* 2007: Interstellar-terrestrial variations; variable cosmic environments, the dynamic heliosphere and their imprints on terrestrial archives and climate. *Space Science Reviews* 127: 327-465

262: Mörner, N.-A. *et al.* 2013: General conclusions regarding the planetary-solar-terrestrial interaction. *Pattern Recogn. Physics* 1: 205-206

263: http://www.southampton.ac.uk/news/2015/05/28-ocean-circulation-study. page

264: http://dailycaller.com/2015/03/03/paper-global-warming-more-like-global-cooling/

265: http:www.eike-klima-energie.eu/climategate-anzeige/waermerbilanz-der-erde-und-globale-temperaturaenderung/

sunspot activity may indicate the end of the recent solar maximum stage. The cycle peaked in April 2015 and the second of the two small peaks at the zenith of this cycle was the stronger.

Sunspot records going back to 1750 AD show that the stronger second sunspot peak only occurs in a long-term declining period of solar activity just prior to a cold period. NASA and the solar physics community agree that Solar Cycle 24 is the weakest in 100 years. The Space and Science Corporation argue that the next cold period has already begun;[266] this will be confirmed if Solar Cycles 25 and 26 are weaker and if the projected colder period will continue until the 2040s.

Between 1755 AD and now, there have been only a few solar cycles that have had fewer sunspots during the maximum phase of a solar cycle. Solar activity is declining, possibly faster than at any time over the last 9,300 years. Habibullo Abdussamatov, director of Russia's space research laboratory, has long been advocating that another Little Ice Age is coming.[267] Maybe the brutal winters in the Northern Hemisphere of 2013-2014, Arctic cooling since 2010, the lack of a temperature rise for more than 18 years and seasonal crop planting delays because of cool wet springs in the 2013, 2014 and 2015 Northern Hemisphere are trying to tell us something. Maybe global warming now no longer exists. Are we starting a cooling phase? Time will tell.

If the atmosphere were to be a couple of degrees warmer, then more CO_2 would be released from the oceans and plant life would thrive. Crops, forests and grasslands would grow faster and better in a longer growing season over larger areas of land that would support more habitats, wildlife, agriculture and people. However, if the planet underwent a solar-driven cooling like the Dalton Minimum

266: http://www.spaceandscience.net/id74.html
267: https://meteolcd.wordpress.com/2009/11/01/abdussamatov-the-sun-defines-the-climate/

(1790-1830 AD) or the Maunder Minimum (1645-1715 AD), agriculture would be reduced in Canada, northern Europe and Russia because of a reduced growing season and cooler weather. Moisture would be stripped out of the atmosphere and, as with all previous cold periods, desertification would be common. There would be less grain for export to the subcontinent, Middle East and North Africa. There might be no excess grain for export. There might not be enough grain to make biofuels.

The ability to feed the world's growing population would be impaired but, even under colder conditions, fertilisers, insecticides, fungicides, GM crops and mechanised farming would still be able to allow maximum food production under cooler conditions. More forests would be converted to croplands and more unpredictable storms and cold snaps during the growing season would make agriculture more hazardous. The Little Ice Age (1350-1850 AD) may be a window into the next cooling. During this time there was crop failure, malnutrition, starvation, disease and war in much of Europe. God help us if we have the next inevitable glaciation. Cold weather kills people at a higher rate than warm weather.

Some groups such as the Space and Science Research Corporation[268] predicted in April 2015 that the historic period of growth in the USA and global agricultural output would decline for decades when the Sun reaches its solar hibernation (which occurs every 206 years). This prediction would be accentuated by the fact that governments, corporations and some farmers are preparing for global warming and have not prepared for a colder period. Cooling is inevitable. It is just a case of when.

In Western countries with insulation and home heating, people in high latitudes can already survive very cold weather. Carbon taxes, renewable energy mandates and subsidised renewable energy have

268: http://www.spaceandscience.net

increased electricity and home heating costs. When energy is costly or rationed, employers lay off workers. With the high cost of energy, we get a glimpse of the future. Poor pensioners in the UK now sit in buses, libraries, shopping centres and other places for much of the day because they can't afford home heating. This is fuel poverty. In Germany, Greece and other countries, the rising energy costs have resulted in people taking their axes to the forests to cut wood for heating and cooking, others burn books in stoves because books are cheaper than wood or coal and some families have resorted to poaching.

The Vatican has operated an astronomical observatory (*Specola Vaticana*) for hundreds of years. It still does. The Gregorian calendar of 1582 AD was a refinement of the Julian calendar. Easter is determined from the lunisolar calendar and does not occur at a fixed calendar date (no matter whether the Julian or Gregorian calendar are used). Easter in Western Christianity falls within seven days of the astronomical full moon on the Gregorian calendar whereas Eastern Christianity uses the Julian calendar and Easter may be later. Dates for the astronomical Easter, Gregorian Easter, Julian Easter and Jewish Passover are different and, in the past, the precise date of Easter has been contentious. The Vatican used its advice from astronomers and mathematicians (including those from *Specola Vaticana*) to set the Gregorian calendar and Easter.

One would have thought that with the Vatican's long interest in astronomy, the Pope would have sought advice from his own astronomers about the effects of the Sun on the Earth's climate. The Vatican Observatory Foundation is very active in astronomy, reports the latest astronomical findings (e.g. NASA's exploration of Pluto)[269] and clearly is abreast of the latest scientific findings. The papal Encyclical seems to only use the IPCC reports as the sole scientific reference.

269: www.vofoundation.org

Maybe the Encyclical makes no mention of the role of the Sun in the Earth's climate because the IPCC dismisses its importance.

When I am asked as a scientist whether the Earth will cool or warm in the future, my answer is yes.

Sea ice

Sea ice has been used for climate scares. In 2007, the media and green left activist scientists were telling us that the Arctic would be free of ice by 2013.[270] It appears that previous projections had underestimated the processes of ice loss and a new high-resolution regional model for the Arctic Ocean and sea ice forced with realistic atmospheric data gave the new estimate.[271] This estimate was supported by other scientists such as Peter Wadhams who stated:[272]

> *Some models have not been taking proper account of the physical processes that go on. The ice is thinning faster than it is shrinking and some modelers have been assuming the ice was a rather thick slab. Wieslaw's model is more efficient because it works with data and it takes account of processes that happen internally in the ice.*

So there you have it. The Arctic is doomed. Wadhams has also claimed that three British climate "scientists" have been murdered by climate skeptics and that he is on a hit list.[273] This poor soul is an advisor to Pope Francis.

The year 2013 has been and gone so this prediction can be evaluated. The ice has expanded. The problem is that in recent years there has been more ice in the world at any time since satellite records began in 1979 and its thickness is increasing. The Arctic summer ice, which was retreating recently, has now started to advance. These are

270: http://news.bbc.co.uk/2/hi/7139797.stm
271: Fall meeting 2007, *American Geophysical Union*; paper by Wieslaw Maslowski
272: http://news.bbc.co.uk/2/hi/7139797.stm
273: *The Times*, 28 September 2015

measurements, not models. And by the way in case you are worried, polar bear numbers are increasing and temperatures in Greenland have shown no increase for decades.

Sea ice (and the continental ice sheets) gives us clues about how climate changes. When an ice block is touched, it feels cold. This is because the ice is taking heat energy from your hand to make it melt. When sea ice melts, it takes heat energy from the ocean to melt. There has been a rapid growth of sea ice in Antarctica to 20 million square kilometres and a recovery of the Arctic sea ice.

Because the latent heat of fusion of ice is 80 calories per gram per °C, as more ice forms from seawater heat is given out to the atmosphere. Because there has been no recorded increase in atmospheric temperature at the time sea ice is increasing, then the heat given out must be compensating for an atmosphere that is cooling. Arctic sea ice area should be related to the regular 60 to 70-year Atlantic Multidecadal Oscillation (AMO)[274] which peaked with its warm cycle in 2012 hence post-2012 we could be facing a more significant decline in global atmospheric temperature as predicted by the total solar intensity.

The US submarine *Nautilus* was able to surface at the North Pole through thinned ice in 1956 and during the previous AMO thinning, Amundsen was able to navigate the Northwest Passage by sail in 1906. Both these times of thinner Arctic sea ice were when the atmospheric CO_2 content was less than at present. Trillions of dollars are being spent in case there is global warming driven by human emissions of CO_2. Maybe we need to worry about the next cycle of cooling that could last for over 100 years.

The Pacific Decadal Oscillation (PDO) is a switch between two circulation patterns that occurs every 30 years. It was originally dis-

274: Andronova, N. G. and Schlesinger, M. E. 2000: Causes of global temperature changes during the 19th and 20th centuries. *Geophys. Res. Lett.* doi: 10,1029/2000GL006109

covered in 1997 in the context of salmon production. The warm phase tends to warm the land masses of the Northern Hemisphere.[275] The AMO and PDO data sets are not similar and cannot be added or averaged. In the 1930s, the AMO and PDO warm phases were coincidental. This was a period of time of record temperatures and dustbowls in the USA. Both cold phases came together in the 1960s and 1970s, the time of scares that the planet was entering another ice age.

For a decade from the mid 1990s, the warm phases again coincided and this was the last time the Earth was warming. The PDO has now turned cold, the AMO has peaked in its warm phase and temperatures are neither increasing nor decreasing. From the mid 2020s to the mid 2030s, both the AMO and PDO will be in their cold phases and planet Earth could experience a similar cooling to that of 1964-1979. Solar activity is declining as well and some solar physicists are predicting a Dalton Minimum or even a Maunder Minimum level. Time to buy a warm coat.

Polar ice is the poster child for the climate catastrophists. When the Dark Ages ended (~900 AD) and the planet started to warm naturally in the Medieval Warming (900 to 1280 AD), the Vikings were the first to feel it. They enlarged their fishing grounds and invaded countries to the south as illegal boat people. Although climate catastrophists ignore the Medieval Warming, their narrative is that if the planet was warming due to human emissions of CO_2, then the poles would be the first to feel it. We all know that, because of our sinful emissions of CO_2, polar ice is retreating at an alarming rate and polar bear numbers are declining.

A good Gaia-fearing decent green left environmental activist would face heroic hardships and go to the poles to get first hand evidence. That would be proof. The deluded demonic denialists would

---

275: http://www.droyspencer.com/global-warming-background-articles/the-pacific-decadal-oscillation/

be finally silenced. Such trips have attempted to view the reduced polar ice and were to regularly report back horror stories. Such trips were so well planned that ideology was all that was needed. To view the Cryostat satellite data[276] and polar weather station temperature records would not be necessary because this was the sort of data that denialists used.

We are told that the area of Arctic sea ice had been decreasing for some years. We were all doomed. However, the European Space Agency[277] reported that the Arctic sea ice had increased by a staggering 33% in 2013 and 2014. Canadian researchers on an icebreaker in Hudson Bay had to suspend their activities because the ice was thicker than it had been for the last 20 years. Weather stations in western Greenland (Godthab Nuuk) and eastern Greenland (Angmagssalik) have measured temperature since 1900. These stations are on opposite sides of Greenland hence give a good coverage of the island. As expected, the unadjusted temperature data show some years are colder or warmer than others but over the 115-year record of measurement, there has been neither a significant warming or cooling of Greenland. This is public record GISS data freely available from NASA and NOAA.[278]

In 2008, Gordon Pugh and two colleagues were sponsored by a climate risk insurance company and backed by Al Gore, the BBC and the Prince of Wales. They left Svalbard (Norway) for a leisurely 1,200 km paddle across the Arctic Ocean to the North Geographic Pole. The intention of the expedition was to electronically measure ice thickness and show how much Arctic ice had vanished. The equipment froze after a few days and a tape measure was used. After a few

---

276: http://www.esa.int
277: http://www.esa.int/Our_Activities/Observing_the_Earth/Cryosat/Arctic/sea-ice-up-from-record-low
278: 1900-2010: http://data.giss.nasa.gov/gistemp/station_data_v2/; 2011-2015: http://data.giss.nasa.gov/gistemp/station_data/

weeks, the courageous expeditioners had to be helicoptered[279] back to a rescue ship because the constantly moving ice was too thick. The expedition was abandoned 135 km from Svalbard. Pugh and his fearless fellow fools couldn't find a gap in the ice despite many tries. If Pugh had used existing data from satellites and Greenland temperature measurements or even been rational, he would never have attempted his ill-fated trip.

Clearly the Englishman Pugh does not watch quality British television. If he did, he would have seen that in 2007, the *Top Gear* crew took 9 days to drive a Toyota Hilux from Resolute Bay (northern Canada) across sea ice to the 1996 Magnetic North Pole (78°35.7'N, 104°11.9W). The ice was so thick that the vehicles did not end up on the bottom of the Arctic Ocean.[280] The North Magnetic Pole is moving northwards and westwards and is currently about 1,600 km from the North Geographic Pole (90°N, 0°E or W).

Pugh has form. In July 2007 when the Arctic sea ice was at its lowest minimum, he swam one kilometre across an open patch of sea at the North Pole without a wet suit in order to bring attention to the environment. It didn't, although his actions might have interested some psychiatrists. He has also had short swims in the seven seas and across the Maldives in order to bring attention to climate change and the health of the oceans. It didn't. In 2010, Pugh swam across a glacial lake near Mt Everest to show that Himalayan glaciers were melting and that the impact of reduced water supply will have

---

279: Apologies for not using a correct passive gerund, you know what helicoptered means anyway.

280: Jeremy Clarkson was filmed driving and drinking gin and tonic. After complaints from temperance-type hand wringers, Clarkson claimed that he was over international waters and therefore sailing. The highest rating television show in the world showed that the Arctic sea ice was not thinning. Greenpeace complained that *"Clarkson is a problem because he has represented some climate sceptic views."* That's the true ecofascist face of Greenpeace, no one is allowed to have a different opinion.

an effect on world peace. Himalayan glaciers melt a little in summer and grow a little in winter.

An attempt at climate alarmist niche summer tourism to Antarctica in January 2014 ended in farce. But it could have been a tragedy with multiple fatalities. It was promoted as an *"expedition to answer questions about how climate change in the frozen continent might already be shifting weather patterns in Australia"* by retracing the steps of Sir Douglas Mawson 100 years earlier. The tourists on this largely taxpayer-funded jaunt that cost $1.5 million found no flowers growing in meadows around Mawson's Hut in Antarctica and were not able to return as heroes with the proof that human-induced global warming is thinning Antarctic sea ice.

Chris Turney, plus wife and children, mustered paying tourists and a sympathetic free-loading media from the BBC, ABC, *Guardian* and Fairfax Press onto the fossil fuel-burning *Akademik Shokalskiy* to watch with bated breath the heroic planet-saving scientists battling against the elements to measure the thinning ice because we all know that human emissions of CO_2 from burning fossil fuels drive global warming and results in melting of polar ice. Never mind the huge amount of fossil fuels burned to make these measurements and to get to Antarctica. No wind- or solar-powered transport was used. The activist "scientists" ignored measurements made far more easily from satellites and history that show that the Antarctic ice sheet is currently expanding. If they had used existing data or been rational, they would never have made the ill-fated trip.

However, nature has a sense of humour. The Russian gin palace was trapped in ice, a fossil fuel-burning Chinese ice breaker sent to rescue these heroic adventurers also became trapped in ice, the real Antarctic research from bases was interrupted as an Australian fossil fuel-burning ice breaker supply ship was diverted and the climate tourists were eventually flown by a fossil fuel-driven helicopter to the warmth of a fossil fuel-heated ship well away from the ice. The

Americans, Australian and Chinese all ran up huge costs to rescue the passengers from the Ship of Fools. All sorts of excuses were invented to show that the climate "science" activists on the ship were not ill prepared, incompetent, ignorant or unaware of past ground and satellite measurements. When questioned about the failure, they resorted to obfuscation and dissimulation.

The climate activist community was silent, the normal suspects in the media became very creative with excuses (especially from those on board) and the journal *Nature* showed that it was a magazine of political activism rather than one of scientific independence. The expedition was to show that this area was warmer than when Sir Douglas Mawson was in the exact same place 100 years ago. The farce showed the exact opposite. Mawson was able to get much closer to the land in his coal-fired steam driven yacht the *Aurora* because of the lack of ice.

There was no chance of frostbite, starvation, eating huskies or death of companions, as Mawson experienced for almost two years. No huts had to be built in gale-force winds. It was all a bit of a giggle for a couple of weeks with a games program organised on the ice because passengers were getting bored while stuck in the ice. Dozens of tourist vessels visit the Antarctic without becoming trapped in ice. They look at satellite data on the extent of the sea ice. It appears that the only tourist ship ever to be trapped in ice in summer was one with climate "scientists" trying to show that the ice was disappearing.

If Chris Turney did not live off research grants and was not employed by a university, he would have been sacked for gross incompetence, breaches of safety protocols and misleading and deceptive conduct. The taxpayer still keeps paying him. However, the public was not fooled.

Both poles have a strange effect on climate activists. There must be something in the organic water[281] that climate activists drink

---

281: A pedantic joke for chemists

whereby they ignore measurements from satellites and ground ob-
servatories and just follow their ideology.

Surely this is the behaviour of religious fundamentalists and not
that of rational people?

Ice sheets and glaciers

The Pope acknowledges some scientific uncertainties yet is in full
flight with depressive disaster dogma when he claims that changes
to the Earth, life, oceans, atmosphere and ice sheets are primarily
a result of human consumption, emissions of CO_2 and waste. For
some 80% of time, the Earth has had neither ice sheets nor sea ice.
For the last 34 million years, Earth has been in an ice age wherein ice
sheets expand during a glaciation and retreat during an interglacial.
Furthermore, the ice sheets expand and contract[282] at the poles for
many reasons, the main ones being seasons, orbit, gravitational surg-
ing of ice flows, sub-glacial melt water flow, air and sea temperature
changes and volcanic activity. Earth is currently in an orbitally-driven
interglacial that should end on a Thursday.

For decades it has been known that ice flow is not by melting
but by creep.[283] Creep may be partially assisted by water acting as a
lubricant at the base of an ice sheet.[284] The ice sheets of Greenland
and Antarctica lie in basins (i.e. with raised edges) hence ice must
flow uphill before it then flows downhill to glaciers that disgorge into
the oceans. A simple process of melting of ice by global warming
entraps melt waters in sub-glacial basins. The processes of ice expansion and contraction are very complicated.

For a long time it has been known that there are at least 20 volca-

---

282: Augustin, L., 2004: Eight glacial cycles from Antarctic ice. *Nature* 429: 623-
628
283: Perutz, M. F. 1940: Mechanism of glacier flow. *Proc. Phys. Soc.* 52: 132-135
284: Stearns, L. A. et al. 2008: Increased flow speed on a large East Antarctic
outlet glacier caused by sub-glacial floods. *Nature Geoscience* 1: 827-831

noes beneath the West Antarctic Ice Sheet,[285,286] volcanoes beneath ice in the Yukon, and British Columbia are well documented[287] and recent data shows that the heat flow beneath the ice sheet has probably been underestimated.[288] Some areas underneath the ice sheet have local volcanic heat sources and others have a widespread area of high heat flow from deep in the Earth. This is not unusual and has happened many times before.[289] The instability of the ice sheet could well be related to variable flows of geothermal heat beneath the ice sheet rather than sea or air temperature changes. Science constantly makes new and surprising discoveries and use of dogmatic words such as consensus or settled science shows that the environmental activism of the Pope and others is unrelated to science.

We are in an ice age with alternating glaciations and interglacials. We know when it started. We don't know when it will end. This ice age was a long time coming. India collided with Asia 50 million years ago. Local climate was changed and CO_2 continued to be drawn down into soils and sediments from the atmosphere as the mountains rose. The Earth's climate has been cooling for the last 50 million years.

South America had the good sense to pull away from Antarctica 37 million years ago, a circum-polar current isolated Antarctica from warm water and local ice caps joined to form a single continental ice sheet on Antarctica some 34 million years ago. The Antarctic ice

285: LeMasurier, W. E. 1976: Intraglacial volcanoes in Marie Byrd Land. *Antarctic Jour. U.S. 11*: 269-270

286: Corr, H. F. J. and Vaughan, D. G. 2008: A recent volcanic eruption beneath the West Antarctic ice sheet. *Nature Geoscience* 1: 122-125

287: Plimer, Ian 2009: *Heaven and Earth. Global warming: The missing science.* Connor Court, p. 279

288: Fisher, A. T. *et al.* 2015: High geothermal heat flux measured below the West Antarctic Ice Sheet. *Sci. Adv.* 1: e1500093, 9pp

289: Steig, E. J. *et al.* 2013: Recent climate and ice-sheet changes in West Antarctica compared with the last 2,000 years. *Nature Geoscience* 6: 372-375

sheet has waxed and waned. At present, we are approaching the end of an interglacial and, unless we can change the behaviour of the Sun and the Earth's orbit, we will enter the next inevitable glaciation.

As planet Earth cooled, the slight cyclical variations in the Earth's orbit and distance from the Sun started to have a profound influence on climate. However, there were some short sharp periods of global warming unrelated to CO_2 or industry (as humans were not around then). Climate changes drove human evolution over the last five million years. In southeastern USA, between 5.2 and 2.6 million years ago, atmospheric CO_2 content was more than now as were global temperatures (2 to 3°C) and sea level (10 to 25 metres). This was probably driven by the regular changes in the Earth's orbit.

With the closure at Panama of the connection between the Atlantic and Pacific Oceans 2.67 million years ago by volcanoes, Earth started to cool. Coincidentally, there was a supernoval eruption at the same time. The bombardment of the Earth by cosmic rays from this supernova eruption led to the formation of low-level clouds that cooled the Earth's surface. The two coincidental processes accelerated cooling. As a result, the Greenland ice sheet formed.

Climate fluctuated in cycles between warm and cold periods every 41,000 years. This was driven by changes in the Earth's axis. About one million years ago, the climate started to fluctuate between cold and warm periods on 100,000-year cycles driven by changes in the orbit from elliptical to circular patterns. We are currently in the warmer interglacial phase of an ice age that has been in progress for 34 million years. During the last interglacial, sea level was four to nine metres higher than now and temperatures were 3 to 5°C higher than now. During the current interglacial 6,000 years ago, sea level was two metres and temperature 2°C higher than now. We cannot escape the fact that the current interglacial will end and we will enjoy another 90,000 years of glaciation. Previous glaciations had kilometre-thick ice sheets that covered Canada, northern USA, most of the UK, most of Europe north of the Alps, most of Russia, all of Scandinavia and elevated areas in both hemispheres. Much of the Andes, New Zealand and Tasmania were covered by ice. There is no reason why the next inevitable glaciation will be any different.

Upland areas, even in the tropics, had glaciers. In areas with no ice sheets, strong cold dry winds shifted sand and devegetation occurred. Dunes in Australia, Asia, North Africa, the Middle East and North America again moved and great wind-deposited loess deposits covered Mongolia, China and northern USA. When Earth eventually has another glaciation, there will be mass depopulation, devegetation, dunes, extinctions, sea level fall and the destruction of coral reefs. It's happened before, it will happen again.

In Australia, the Great Barrier Reef, the poster child of the green left environmental activists, disappeared during glacial events more than 60 times over the last three million years. It reappeared after every one of these events. The Great Barrier Reef first formed about 50 million years ago and has survived hundreds of coolings and warmings and massive rain events that deposit sediment on the Reef. The sea level fall and lower temperature during glacial events kills higher latitude coral reefs and they continue to thrive at lower latitudes. The geological record shows that coral reefs love it warm, especially when there is more CO_2 in the atmosphere. During glaciation events, existing data shows tropical vegetation is reduced from rainforest to grasslands with copses of trees, somewhat similar to the modern dry tropics inland from the Great Barrier Reef. There was no Amazonian rainforest during the last glaciation.

Humans have adapted to live by the sea, in the mountains, in deserts, on ice sheets, in the tropics and in the artificial reality of cities. In warm times, history shows us that humans thrived. The biggest stresses humans endured from climate were the changes from interglacials to glaciation.

History shows that we have numerous extreme weather events

during times of cooling, not warming. We are now in an interglacial facing the next inevitable glaciation. Humans have shown many
times in the past that they can easily adapt to warming. We saw from
the Minoan, Roman, Medieval and Modern Warmings that life is far
better in warm times than in cold times. We now have technology
that makes it easier to adapt than in former times. If planet Earth
is warming, then why is warming such a problem and how can a tax
make warming go away?

I fear the Pope has fallen into the trap set by his atheistic activist
advisors of providing a simple answer to a complex natural phenomenon.[290] The science is far from settled and for the Pope and activist
scientists to suggest that there is a consensus or that the science is
settled[291] show that minds are closed and he is receiving poor advice
from activist scientists.

Warming

We are told that a slight warming of the Earth's atmosphere will create all sorts of disastrous environmental problems. Extreme weather,
extinctions, sea level rise, destruction of coral atolls and ocean acidification are the current scare. Hang around a bit longer and I am
sure new scares will be concocted just for you. A warming of 2 to
5°C would only bring the Earth back to where it was in the Medieval Warming 1,000 years ago, the Roman Warming 2,000 years ago,
the Minoan Warming 3,000 years ago, the Egyptian Old Kingdom
Warming 4,000 years ago and numerous other warmings including
the Holocene Optimum which peaked 6,000 years ago.[292] Some of
the intervening colder spells were colder than the Little Ice Age.

290: Misquote from H. L. Mencken in "The Divine Afflatus" (*New York Evening
Mail*, 16 November 1917): For every complex problem there is a solution that is
neat, simple and wrong.
291: *Laudato Si'*, Paragraphs 23, 164 and 165
292: Alley, R. B. 2000: The Younger Dryas cold interval as viewed from central
Greenland. *Quat. Sci. Rev.* 19: 213-226

Extreme weather

There is no correlation between extreme weather, increasing CO_2 levels and temperature. It is the opposite. This has been shown from the record of northern Australian tropical cyclones.[293] The same has been shown from the other side of the world. Between 1860 and 1875, storminess was extremely high in northern Denmark. Since then, it has decreased. This validates what has been seen at a number of measuring stations in NW Europe.[294]

Each time there is a hurricane in the US, there are suggestions that this is due to global warming. Much of the hurricane damage in recent times is because Americans are wealthier and have built palaces along the coast. Furthermore, some coastal cities are subsiding, rather like Venice (Italy). Data tells a different story. The annual intensity of land-falling hurricanes from 1900 to 2012 has not changed and, if anything, has slightly decreased. Big storms made the land in 1904, 1924, 1925 and 2012.[295]

Extinction

For centuries, human beings have been causing other species to become extinct at a rate higher than the normal species turnover rate. There are apocryphal suggestions that Earth is entering its sixth mass extinction of multicellular life.[296,297,298] This is misleading as the

293: Dowdy, A. J. 2014: A three-decade history of Australian region tropical cyclones. *Atmos. Sci. Lett.* 15: 292-298

294: Clemmensen, L. B. *et al.* 2014: Storminess variation at Skagen, northern Denmark since AD 1860: Relation to climate change and implications for coastal dunes. *Aeolian Res.* 15: 101-112

295: http://wattsupwiththat.com/2012/11/27/an-update-to-us-hurricane-intensity-1900-2012-no-recent-trend-with-hurricane-sandy/

296: Pimm, S. and Raven, P. 2000: Biodiversity: Extinction by numbers. *Nature* 403: 843-845

297: Barnosky, A. *et al.* 2011: Has the Earth's sixth mass extinction already arrived? *Nature* 471: 51-57

298: Ceballos, G. *et al.* 2015: Accelerated modern human-induced species losses: Entering the sixth mass extinction. *Environ. Sci.* 1: doi 10.1126/sciadv.1400253

Earth has had five major mass extinctions and many minor mass extinctions. There are the normal scientific problems in ascertaining the number of species that were present at any one time on Earth because not all species become fossilised and new fossils are being found every day.

Some 60-70% of all species disappeared in the Ordovician-Silurian mass extinction events (450-440 Ma), 70% in the Late Devonian mass extinction (375-360 Ma), 90-96% in the Permian-Triassic mass extinction event (251 Ma), 70-75% in the Triassic-Jurassic mass extinction event (201 Ma) and 75% in the Cretaceous-Tertiary mass extinction event (65 Ma).[299]

Minor mass extinction events occurred at 2,400 Ma, 542 Ma, 517 Ma, 502 Ma, 488 Ma, 428 Ma, 424 Ma, 420 Ma, 416 Ma, 305 Ma, 270 Ma, 232 Ma, 183 Ma, 145 Ma, 117 Ma, 33.9 Ma, 15.5 Ma, 2 Ma, 0.64 Ma, 0.074 Ma and 0.0013 Ma. Get used to it: Extinction is normal and occurs for a great diversity of reasons. No species, including *Homo sapiens*, is on Earth forever. When we shuffle off, we vacate an ecosystem for another species to thrive. Not one previous mass or minor extinction event was due to global warming. We are certainly seeing a loss of species, this could be by normal species turnover but it is certainly not at the levels of a mass extinction, whether a major or minor mass extinction.

Whether this is the sixth major mass extinction, a minor mass extinction or an accelerated turnover of species is debatable. The Pope suggests that we humans are responsible for extinctions due to agriculture,[300] infrastructure and habitat loss[301] and global warm-

---

299: Raup, David 1992: *Extinction: Bad genes or bad luck?* W.W Norton

300: *Laudato Si'*, Paragraph 24

301: *Laudato Si'*, Paragraph 35

ing.[302] He comments on the extinction of mammals and birds.[303] The popular view is that current extinctions are due to habitat loss and climate change. One of the main causes of extinctions has been ignored. It is invasive species.[304] The Pope does not mention invasive species. Waves of extinctions occurred in the Caribbean in the 1500s, the extinction rate rose again during the first phase of expansion and exploration in the 1700s and then rose again during the age of empires after 1850 that peaked at the beginning of the 20th century. There is now no *Terra Incognita* that hasn't been visited by explorers and hence the worst of the extinctions from introduced predators are behind us.[305]

Predators, parasites and pests were brought in by explorers and settlers and have had a devastating effect, especially on islands. We really don't know how many species of life exist on Earth. New species are continually being described. Although 1.5 million species have been discovered, described and classified, there may be anything from 2 million to 50 million species on Earth with one source claiming that there are 8.7 million species.[306] This list does not include microorganisms that constitute the greatest biomass on Earth. By contrast, there are only 5,026 valid mineral species[307] approved by the International Mineralogical Association.

302: *Laudato Si'*, Paragraph 24

303: *Laudato Si'*, Paragraph 34

304: Matt Ridley, *The Times*, 22 June 2015

305: http://wattsupwiththat.com/2013/07/07alexander-the-great-explains-the-drop-in-extinctions/

306: http://www.nature.com/news/2011/110823/full/news.2011.498.html

307: Including plimerite

Table 1: Estimated number of animal and plant species on Earth[308]

Group	Number of species
Vertebrates	
Amphibians	6,199
Birds	9,956
Fish	~30,000
Mammals	5,416
Reptiles	8,240
Subtotal	59,811
Invertebrates	
Insects	~950,0001
Molluscs	~81,000
Crustaceans	~40,000
Corals	2,175
Others	~130,000
Subtotal	≥1,203,375
Plants	
Mosses	~15,000
Ferns	13,025
Gymnosperms	980
Dicotyledons	199,350
Monocotyledons	~59,300
Green algae	3,715
Red algae	5,956
Subtotal	≥297,326
Others	
Lichens	~10,000
Mushrooms	~16,000
Brown algae	2,849
Subtotal	≥28,849
TOTAL	≥1,589,361

308: http://www.factmonster.com/ipka/A0934288.html

Some 77 mammal species (out of a total described species count of over 5,000), 34 amphibians (out of over 6,000 species) and 140 bird species (out of a total of about 10,000 described species) have become extinct in the last 500 years.[309] There may be others but it is the order of magnitude that is important. Europe has lost only one bird species over the last 500 years.

Many other species may be on the brink of extinction and are sitting in the waiting room (e.g. South China tiger, Sumatran elephant, Amur leopard, Atlantic goliath groper, Gulf porpoise, Northern bald ibis, Hawksbill turtle, black rhinoceros, pygmy three-toothed sloth, Chinese pangolin etc). Great efforts are being made by conservationists and zoos to breed threatened species and some species that were thought to be extinct are sometimes rediscovered later (Coelacanth, Bermuda petrel, Chacoan peccary, Lord Howe Island stick insect, Monito del Monte, La Palma giant lizard, takahe, Cuban solenodon, New Caledonian crested gecko, New Holland mouse, giant Palouse earthworm, large-billed reed warbler, Laotian rock rat etc). There are clearly many more Lazarus species to be found.

Of the known total birds and mammals that have become extinct almost all lived on islands with only nine that lived on a continent (excluding the island continent Australia).[310] Island species that became extinct over the last 500 years included the dodo, Steller's sea cow, the Falklands Island wolf, the quagga, the Formosan clouded leopard, the Atlas bear, the Caspian tiger and the Cape lion. On a per unit area basis, the extinction rate on islands was 177 times higher for mammals and 187 times higher for birds than on continents. The continental mammal extinction rate was 0.89 to 7.4 times the background rate whereas on islands it was 82 to 702 times the background rate. Continental bird extinction rate was 0.69 to 5.9 times

309: International Union for Conservation of Nature; www.iucn.org
310: Loehle, C. and Eschenbach, W. 2011: Historical continental bird and mammal extinction rates. *Diversity Distrib.* doi: 10.1111/j.1472-4642

the background rate and on islands it was 98 to 844 times background rate. Fossil assemblages on islands not only show dwarfism but also accelerated extinction. The increased island extinction rate is mainly due to the introduction of predators (including man), microorganisms and diseases.

The carked continental critters are the Bluebuck antelope, Algerian gazelle, Omilteme cottontail rabbit, Labrador duck, Carolina parakeet, slender-billed grackle, passenger pigeon, Colombian grebe and Atitlan grebe. This extinction rate is not apocalyptic, as we have been led to believe by doomsdayers. To call it a mass extinction is an over-exaggeration.

The invasion of hunter-gatherer humans with stone-tipped spears across the Bering Strait into North America 12,000 years ago led to the extinction of many grassland macrofauna species. Some threatened species are now being revived. Explorers and settlers brought rats, sheep, goats, cats, dogs, pigs, mosquitoes and avian malaria onto Hawaii, through the Caribbean, the South Atlantic islands, the Indian Ocean islands and the Pacific islands. Young mammals, birds, reptiles and amphibians have been eaten by introduced animals, especially rats and cats. Hawaii has lost about 70 species of bird since Captain Cook was killed on his third visit to Hawaii on Valentine's Day, 1779.

If the green left environmental activists really wanted to make Australia a better place, they would eradicate every introduced feral animal with guns, traps, poisons, genetic engineering, organism specific viruses and pesticides. In Australia, feral cats, foxes and rats kill native birds, mammals, amphibians and reptiles. Feral goats just eat everything in their path. Feral dogs form marauding killer packs that have even killed children, feral horses destroy alpine upland areas in national parks, cane toads wipe out wild life, pigs pollute and destroy waterways, camels kill vegetation in sensitive desert areas, introduced fish replace native fish and introduced plants dominate some native plant habitats.

The time has come to close our ears to green left environmental activist noise and to watch them killing introduced plants and animals. Where are the green left environmental activist armies in rural and outback Australia conducting genocide of introduced species? Time to leave the comforts of a city life made possible by low cost coal-fired electricity and actually do something to make the world a better place. Don't hold your breath, it is easier to criticise from a city media studio than to actually do something useful.

In the UK, introduced species such as grey squirrels, mink and signal crayfish have already pushed the endemic red squirrel, water voles and native crayfish populations towards extinction. British native animal populations are decreasing because of ash dieback and introduced zebra mussels, harlequin ladybirds, Chinese mitten crabs, New Zealand flatworms and muntjac.

To argue that economic development should be reduced or stopped because of extinctions is flawed. It is also flawed to spend billions on climate change because climate change may create extinctions. Introduced organisms create local extinctions. A few decades ago, scientists began noticing a decline in frog and toad populations in Central America. A good example is Costa Rica's golden toad that lived in the cloud forest. It was touted as an obvious example of extinction due to climate change. After all the hysteria died down, the real reason was found. It was an introduced African chytrid fungus. And how did it get to Central America? From the African clawed toad which is used by scientists for laboratory tests.

The Pope claims that we have plundered the Earth and this is

> *reflected in the symptoms of sickness evident in the soil, in the water, in the air and in all forms of life. This is why the earth herself, burdened and laid waste, is among the most abandoned and maltreated of our poor* ...[311]

311: *Laudato Si'*, Paragraph 2

What emotive rubbish. Where is the evidence? Despite the Encyclical having many theological references, there are certainly no
validated or scientific references given as evidence to support this
statement. What really is sickness of the soil, air, water and life? This
is the language of an unbalanced atheistic environmentalist. The
facts are different.

Even in a developed desert country such as Australia where apparently the land is *"burdened and laid waste"*, some 55.8% of the land
is used for low impact grazing, conservation and natural environments comprise 36.7%, cropping 3.5%, forestry 1.8% and intensive
uses such as cities, roads, railways, houses, office blocks, factories
quarries, mines and oil/gas production facilities only occupy 0.4%
of the land area.[312] The intense use of land is very efficient. For example, in Victoria the Minerals Council states that both farming and
mining each contribute to about 9% of the state's GDP, farming uses
70% of the land whereas mining only uses 0.07% of the land area.

In NSW, mining uses just 0.1% of the land compared to 76% for
agriculture, 7.6% for conservation and national parks and 1.8% for
homes and urban development. In NSW, the coal industry directly employs 34,000 people hence with the flow on multiplier, about
140,000 people from mechanics to manufacturers and from hospital-
ity to health professionals.[313] Coal also produces more than 80% of
the state's electricity needs. The Earth is feeding more people from
less arable land than ever in the history of humans.

Sea level

Since the peak of the last glaciation 20,000 years ago, global sea level
has risen 120 metres to a peak known as the Holocene Optimum
(9,000 to 6,000 years ago). Average sea level was two metres higher
and global temperature about 2°C warmer than now. Only a few

312: ABARES, 2010
313: Stephen Galilee, New South Wales Minerals Council, 21 July 2015

thousand years ago, Sydney and Brisbane airports were tidal man-
grove swamps. Sea level always rises after a glaciation. This was the
peak of the interglacial that we now enjoy. It was only 8,000 years
ago that there was no summer ice in the Arctic. The rate of sea level
rise is very variable and is correlated with neither temperature nor atmospheric CO_2. *Nature Geoscience*[314] recently reported that since 2002,
the rate of sea level rise has declined by 31%.

It is a very long bow to argue that human emissions of CO_2 create
global warming which then results in melting of ice and expansion
of seawater to give an increased sea level rise. During the 20th century
when human emissions of CO_2 greatly increased, there were both
warming and cooling periods.

Coastal planning based on global sea level rise ignores local com-
paction, sedimentation, uplift and subsidence. Not only does the sea
level rise and fall, the land level also rises and falls due to tectonic
forces. Many of the rises and falls of both sea and land level are
local, others can be regional and others can be global. Computer-
modelled sea level projections by the IPCC and governments have
already been shown to be wrong. Since the release of the 2007 IPCC
report, the rate of sea level rise predicted by the computer models
has been shown by measurements to be 25% too high.

Areas covered with ice sheets during the last glaciation (116,000
to 11,500 years ago) were pushed down under the weight of ice.
Over time, rocks are plastic and will bend. With the collapse of the
ice sheets in the current interglacial, some lands are rising (e.g. Scandinavia, Scotland) and others are sinking (south-east England, The
Netherlands). Oulu (Finland) is rising at 6.2 mm per year and this
rise is used as a correction factor for IPCC sea level changes. Tide
gauges show that Galveston (Texas) is sinking at 6.6 millimetres per
year, which makes it look as if sea level is rising by 6.6 millimetres

---

314: Milne, G. A. *et al.* 2009: Identifying the causes of sea level change. *Nature Geoscience* 2: 471-478

per year. The IPCC does not correct for this subsidence and only corrects for rising land that was pushed down by the weight of ice. Selectively correcting measurements makes it look as if sea level is rising quickly.

History shows us that some port cities (e.g. Efeses, Turkey) are now inland whereas other cities (e.g. Simena, Turkey) are submerged. These two ancient cities are close to each other and have risen and fallen due to tectonic forces such as earthquakes. In both the Maldives and eastern Australia, relative sea level has fallen. The Maldives is 70 centimetres higher now than in the 1970s and eastern Australia is two metres higher than 4,000 years ago.[315] In the past 2,000 years, sea level has oscillated with five peaks reaching 0.6 to 1.2 metres above the present level.

There is a Commission on Sea Level and Coastal Evolution, an independent body unrelated to the IPCC. This is a group that actually measures sea level, rather than making computer predictions based on incomplete information. It states that by 2100 AD, sea level will have risen by five centimetres with an accuracy of plus or minus 15 centimetres. This means that as a result of their comprehensive specialist studies, they cannot determine whether sea level will rise, fall or be static for the rest of the century as the uncertainty of their measurement is far greater than the actual measurement.

Without a detailed knowledge of local land rises and falls, subsidence, erosion and sedimentation, global sea level predictions for coastal planning are only unfounded speculation. Sea level also rises and falls during El Niño/La Niña events. Sea level can be displaced creating a sea level rise if a large submarine super volcano mass lifts or flows onto the sea floor.

Very detailed global sea level curves for the last 520 million years

---

315: Axel-Morner, N. 2012: *Sea level is not rising.* SPPI Reprint Series have been constructed by the oil industry[316] and researchers.[317] These show sea level rising and falling more than 400 metres. The sequence of sedimentary rocks preserves sea level changes.[318] These changes are related to climate and the shape of the ocean basins which regularly change. These curves are used to ascertain when rising sea water was covering continents,[319] a vital factor for exploration for oil. At present, the sea level is the lowest it has been for 500 million years.

Governments may ask the simple question: How much do we need to spend to cope with sea level rise? This assumes that the sea level is rising in their area, which is commonly not the case. For example, how much should we spend to prevent $4 trillion of Florida real estate being destroyed in the next 500 years? At a rate of discount for the future of 6% per year, the answer is 15 cents! Economic theory tells us that we cannot control what may or may not happen in 500 years time. Common sense says: why bother? Geology says that there may actually be a sea level fall or a sea level rise. The level of uncertainty over a 500-year period is so high that we can equally as well argue that the climate may actually cool and sea level may drop.

The sea surface is not as level as one might expect. High and low pressure meteorological systems have an effect on sea level. For example, there was a huge hump in the South Pacific Ocean extending over an area larger than Australia from October 2009 to January 2010. Wind drove water into an area of an unusually stable high-pressure system. Off the coast of India, the sea level is considerably

316: Vail, P. and Sangree, J. B. 1988: Global sea-level and the stratigraphic record. Sequence Stratigraphy Workbook, Townsville Workshop, 23-25 August 1988
317: Haq, B. U. *et al.* 1987: Chronology of fluctuating sea levels since the Triassic. *Science* 235: 1156-1167
318: Carter, R. M. 1998: Two models: global seas-level change and sequence stratigraphy architecture. *Sed. Geol.* 122: 23-36
319: Marine transgression

lower than the present global average because gravity is weak in one area. Near the coast of Greenland and the Andes Mountains, sea level is higher because of the gravitational pull of mountains. Ocean water sloshes between the west coast of Africa and the east coast of the Americas changing local sea levels.

Coastal areas are very sensitive to changes in climate and sea levels. In Western Australia during the last interglacial 125,000 years ago, there were stag horn coral fringing reefs at Fairbridge Bluff on Rottnest Island offshore from Perth.[320] Sea level was at least 3 metres higher than at present and the seas were far warmer. There were no human industrial emissions of CO_2 125,000 years ago. In the current interglacial, no such coral grows on Rottnest Island and the nearest stag horn corals are 500 kilometres north at Houtman Abrolhos. This suggests that the current interglacial is cooler than the previous interglacial. This is supported by temperature proxy measurements on ice cores.

In polar regions, some suggest that during the last interglacial period, sea level was four to nine metres higher than now and temperature was 3 to 5°C higher than now. Local conditions are such that there can be no definitive figure except that sea level was higher than now and it was warmer. Near King Sound (Derby, Western Australia), there are thickly vegetated sand dunes. These extend as fossil dunes beneath the coastal mud flats and show that dune formation took place 20,000 years ago at the peak of the last glaciation. Northern Western Australia was then arid, cold and windy, sea level was 100 metres lower than now and there was no vegetation on shifting dunes. Since then, the climate has become tropical, sea level has risen and tidal mudflats have covered what were treeless dunes.

320: Baker R. G. *et al.* 2005: An oscillating Holocene sea level. Revisiting Rottnest Island, Western Australia and the Fairbridge Eustatic hypothesis. *Jour. Coast. Res.* 42: 3-14

Not all sea level change is due to climate. Some 97% of ground-water extracted for human use ultimately ends up in the oceans. Some 15% of the observed sea level rise comes from pumped groundwater used by humans. This water eventually flows into the oceans.[321] Not all land level changes are natural. Extraction of groundwater, oil and gas results in compression of sediments and sedimentary rocks and earth tremors are commonly associated. As a result many places are sinking as a result of human activities (e.g. Bangkok, Mexico City, Venice 1920s to 1970s). Vibration and the weight of cities also re-sults in subsidence.

We are told that global sea levels have risen about 1.7 millimetres per year in the 20th century. This, it appears, is a result of us putting CO_2 into the atmosphere. This, we are told, warms the planet, ex-pands the volume of the oceans, melts polar ice and adds water to the oceans. However, since the last glaciation 11,500 years ago, sea level has risen several metres higher than at present then has gradu-ally fallen over the past 6,000 years. We cannot claim that the small amount of sea level rise in our own life is our fault and the rest was natural. The question that has not been answered is: Which part of the recent sea level rise was from human activities and which part was from the normal post-glacial sea level rise?

Some 200 years ago, convicts engraved tide marks on rocks in Tasmania. These show no change in sea level since then. Tide gauge readings since the 1930s in the USA show no increase in the rate of sea level rise. The sea level measurements by satellite shows a rising trend. The problem with satellites is that they don't actually measure sea level. They measure gravity and computers use the observations of gravity to calculate a theoretical even surface of the Earth and from this there are various ways computers calculate sea level. Very

321: http://www.nature.com/news/source-found-for-missing-water-in-sea-level-rise-1.10676

small changes in the computer code, sea floor geology and database can produce apparently rising seas, falling or a static sea levels. According to the GRACE gravitational recovery satellites, sea level actually fell from 2003 to 2008. The ENVISAT estimates that sea level rose at a mean rate of 3.3 cm per century.

There are 159 tide gauge measuring stations around the world used by NOAA to compute sea level change. Many of these are sited on areas where subsidence occurs as a result of sinking piers, vibration and traffic. To get consistency of measurement, tide gauges require regular high precision surveying of the gauge position.

A good example is Port Adelaide where a surveying over time has shown that the tide measuring station is sinking.[322] This is recorded as a sea level rise. Only 68 of the 159 measuring stations used by NOAA are reliable and these show a sea level rise in the order of 1 mm per year. The gold standard reference is in Hong Kong Harbour, which shows a sea level rise of 2.3 mm a year.

This is not in agreement with four nearby tide gauges suggesting that the reference tide gauge is sinking and has been chosen to exaggerate sea level rise. Furthermore, a total of 11 tide gauges at Hong Kong are administered by the Hydrographic Office of the Marine Department (four gauges), the Hong Kong Observatory (six gauges) and the Airport Authority (one gauge).[323] In 2002, the six Hong Kong Observatory tidal measuring stations showed an annual sea level increase of 1.9 ± 0.4 mm per year[324] and NOAA have chosen the maximum possible sea level rise of 2.3 mm per year, again allowing exaggeration.

322: Belperio, A. P. 1993: Land subsidence and sea level in the Port Adelaide estuary: implications for monitoring the green house effect. *Aust. Jour. Earth Sci.* 40: 359-368

323: http://www.hko.gov.hk/tide.marine/hko.htm

324: Ding, X. L. *et al.* 2002: Sea level change in Hong Kong from tide gauge records. *Jour. Geospatial Eng.* 4: 41-49

Some pretty scary predictions have been made about sea level rise. In 2006 Australia's great and glorious climate commissioner Tim Flannery stated:

> *Picture an eight-storey building by a beach, then imagine waves lapping its roof. So anyone with a coastal view from their bedroom window or kitchen window is likely to lose their house as a result of that change.*

At the time Flannery made this prediction, he had a waterfront house. Why hasn't our climate commissioner moved to higher ground? Is it because he does not believe what he says? When I see Flannery move to higher ground, stop travelling in CO_2 emitting jet aeroplanes and vehicles, have meetings in Canberra[325] bureaucrats' offices with no central heating, no lights and no air conditioning, stop using coal-generated electricity for broadcasting, live in a cave with no electricity and live as he preaches, then I might actually seriously consider one of his numerous scary predictions. If coastal real estate loses value and sea level does not rise, can Flannery be sued? Does Flannery want us to lower our standard of living while he lives in the land of luxury? Maybe Flannery and his green left environmental activist mates are only concerned about controlling your life? If the great predictor Flannery wants a long life, he should predict that his early demise is imminent.

Who else has a waterfront home? Kevin Rudd? Yes. Al Gore? Yes. The climate industry seems to have a large mortgage on hypocrisy. The Greens Party politicians flit around business-class in planes pumping huge amounts of CO_2 into the atmosphere and then tell us that we should reduce our emissions or carbon footprint (whatever that might be). The proponents of a "Carbon Tax" and all those in the climate industry that travel to international conferences and doomsday prediction meetings are too willing to tell us to reduce our standard of living.

325: An old Australian bushie's word that means "place of unlimited entitlement"

Coral atolls

An analysis of more than 600 coral reef islands in the Pacific and Indian Oceans shows that some have remained stable (40%) or increased in size (40%). Only 20% have decreased in area yet it is widely assumed in environmental circles that coral islands, atolls and reefs are disappearing with sea level rise. Some islands grew as much as 5.6 hectares in a decade. Tuvalu's main atoll, Funafuti, comprising 33 islands around the rim of a lagoon gained 32 hectares during the last 115 years.[326]

Contrary to popular disaster stories in the media and promoted by local politicians and green left environmentalists, the Pacific Ocean island atoll states are not disappearing due to sea level rise. They are getting larger. This is not new news. We have known for nearly 200 years that coral atolls increase in size with a relative sea level rise.[327] Over the period from 12,000 to 6,000 years ago during the 130-metre post-glacial sea level rise, coral reefs kept up with sea level rise.[328] The coral sand atoll islands were actually produced by the destruction of reef material during the 2-metre sea level fall since the Holocene Optimum. Corals don't have a problem with sea level rise, can adapt to warmer seas and just grow faster. They die when sea level falls. Why didn't the Pope's scientists know this?

In South Tarawa, Kiribati's 15 square kilometre island capital, crowded with some 50,000 people, coral blocks are used for seawalls, causeways between islands and creating new land. This has led to

---

326: Webb, A. P. and Kench, P. S. 2010: The dynamic response of reef islands to sea-level rise: Evidence from multi-decadal analysis of island change in the Central Pacific. *Global Planet. Change* 72: 234-246

327: Darwin, Charles 1842: *The structure and distribution of coral reefs. Being the first part of the geology of the voyage of the Beagle under the command of Capt. Fitzroy, R.N. Smith*, Elder and Co

328: van Woesik, R. *et al.* 2015: Keep up or drown: adjustment of western Pacific coral reefs to sea-level rise in the 21st century. *Roy. Soc. Open Sci.* doi: 10.1098/rsos.150181

greater storm erosion, changes in sedimentation patterns and more common inundation during storm surges. The real danger to coral reefs and atolls is sea level fall and human activity such as removal of coral sand for cement, building of roads and airstrips together with ground water extraction, blasting of reefs for shipping lanes and use of reef blocks or coral sand concrete for sea walls. There have been claims that residents from Kiribati may be the first climate refugees in the world.

In July 2015, the Supreme Court of New Zealand declined an appeal by Ioane Teitiota to overturn a decision not to grant him refugee status because he argued that he was a climate refugee.[329] The claim stated that Kiribati was the world's lowest lying nation and was unsafe due to rising sea levels. Teitiota was on a work visa, which expired in 2010 and claimed that his family would face passive persecution if forced to return to Kiribati because its government is unable to protect them from rising sea levels caused by climate change. The court rejected the appeal not on logic or science but on the basis that Teitiota did not fit the definition of a refugee under the International Refugee Convention. This saved the court from stating the obvious and leaving the door open for bleeding hearts.

Some environmentalists and church figures may mean well by trying to help Pacific islanders but false claims and telling half-truths to promote a scare campaign does more harm than good and does not provide a solution. The Pope's advisors should have steered him away from making uninformed comments about sea level rise[330] and the constant rate of sea level rise.[331] Sea level change is not constant and the Pope's advisors didn't tell him that sea level goes up and down as does the land level.

329: http://www.3news.co.nz/nznews/climate-change-refugees-case-heads-to-supreme-court-2015021117#axzz3ij7jeqfu
330: *Laudato Si'*, Paragraph 48
331: *Laudato Si'*, Paragraph 23

Reefs have been around for a very long time. About 3,500 million years ago, calcareous organo-sedimentary reef-like structures called microbialites appeared. They comprise calcium and magnesium carbonates with the calcium, magnesium and CO_2 extracted from seawater. For the next 2,500 years, photosynthesising blue-green algae (stromatolites) produced shallow marine reefs. They still do (e.g. Lee Stocking Island, Bahamas Banks; Shark Bay, Western Australia). The history of the Earth shows us that stromatolite reefs thrive in warm times, especially when there is a high atmospheric CO_2 content.

Coral reef communities started to appear about 600 million years ago and from about 600 to 540 million years ago, sponge-like animals (Archaeocyathids), stromatolites, calcareous cyanobacteria and algae were the main organisms in reef communities. From about 540 to 350 million years ago, reefs were dominated by complex communities of algae-sponge-coral associations. Some of the corals present in these assemblages are now extinct (e.g. rugosa).

From about 350 to 220 million years ago, the reef assemblage became even more complex and was dominated by algae-bryzoan-coral communities with minor foraminifera, sponges, stromatoporoids and rudist bivales. For the last 220 million years, reefs comprised dominant scleractinian corals as the main reef builder and modern reefs represent the most complex and developed scleractinian reefs in the history of the planet.

Over the history of time, reefs have mainly died out because of sea level fall or coral-fringed islands and the continental shelf have risen above sea level. It is very rare for reefs to die because of sea level rise. Seawater temperature change[332] (especially cooling) and changes in salinity, dissolved oxygen and stability of dissolved chemicals may also kill reefs. Some reefs are killed by inundation by sedi-

---

332: The Global Coral Monitoring Network in 2000 reported that 16% of the world's coral reefs were "effectively lost" in nine months during the 1997-1998 El Niño. The reefs recovered.

ment as a result of flooding in the hinterland. In more modern times, the greatest threats to reefs have been tourists, fishing with cyanide and dynamite, mining of coral for roads, cement and construction, dredging, introduction of competing non-native species, runoff of sewage, nitrates and phosphates, and increased natural and anthropogenic sedimentation rates.

The Great Barrier Reef of Australia is 2,300 km in length comprising 3,000 coral reefs, 600 continental islands and 300 coral cays. It has over 600 species of soft and hard corals and a breathtaking range of other species.[333] The reef has migrated eastwards and northwards and even disappeared when sea level dropped during glaciation. Sea level was 130 metres lower at the end of the last glaciation 11,500 years ago and the Great Barrier Reef did not exist at that time. Rainforests during the last glaciation comprised copses of trees and grasslands (e.g. Amazon Basin).

For the last 3,500 years, there have been reefs in shallow marine settings. Some rare isolated corals live in deep cooler waters today. The fact that reefs existed in previous times when it was warmer and the atmosphere had a higher CO_2 content, shows that modern coral reefs are in no danger if the atmosphere warms and the CO_2 content increases. In fact, it is the exact opposite. The history of time shows us that reefs, be they algal or coral, thrive when it is warmer and there is a high atmospheric CO_2 content. Can the green left environmental activists please show me why the opposite should be true when we have 3,500 million years of evidence written in stone?

Ocean "acidification"

This is a complete furphy. To use the word "acid" with reference to the oceans is fraudulent. The oceans have been alkaline for thousands of millions of years at times when there was far more CO_2 in the atmo-

---

333: http://www.gbrmpa.gov.au/about-the-reef/facts-about-the-great-barrier-reef

sphere than at present. There is no reason CO_2 solubility in seawater would suddenly disobey all the rules of chemistry because we are emitting traces of a trace gas into the atmosphere.

Dissolving CO_2 in the oceans

Carbon dioxide dissolves in seawater. About 70% of the planet's surface is covered by water. Solubility of CO_2 in water is mainly a function of temperature, pressure and salinity.[334] The colder the water, the more CO_2 dissolves.[335] The long-term global average sea surface temperature is 15°C and at this temperature, seawater can dissolve its own volume of CO_2. At 10°C, seawater dissolves 19% more CO_2 than at 15°C and at 20°C seawater dissolves 12% less than at 15°C.[336] Not surprisingly, there is a correlation between the sea surface temperature and the atmospheric CO_2 content. The more saline the water, the more CO_2 dissolves. The higher the atmospheric CO_2 content, the more CO_2 dissolves in seawater.

The oceans continually remove CO_2 from the atmosphere, this CO_2 is used by organisms to make $CaCO_3$ shells of aragonite and calcite. When organisms die, these shells accumulate on the sea floor and later become limestone. The solid rock limestone contains 40,000 times more CO_2 than the atmosphere. Soils contain carbon-bearing organic compounds and carbonate and, upon erosion, the carbon compounds and carbonates may end up as marine sediments. The wind also pumps CO_2 into sea water.[337] In certain circumstances, $CaCO_3$ precipitates on the seafloor.

334: Harned, H. S. and Davis, R. 1943: The ionization constant of carbonic acid in water and the solubility of carbon dioxide in water and aqueous salt solutions from 0 to 50°C. *Jour. Amer. Chem. Soc.* 65: 2030-2037

335: This is inverse solubility. With many substances (e.g. sugar, salt), the warmer the water the more solid dissolves in water.

336: Endersbee, L. 2008: Carbon dioxide and the oceans. *Focus* 151: 20-21

337: Smith, S. D. and Jones, E. P. 1985: Evidence of wind-pumping of air-sea gas exchange based on direct measurements of CO2 fluxes. *Jour. Geophys. Res.* 90: 869-875

In polar areas, the cold surface water absorbs more CO_2 than at the tropics. The cool, dense, high salinity polar water sinks and is carried to lower latitudes where it upwells and releases CO_2 as the water warms. About 70% of ocean degassing occurs by this process (thermally driven solubility pump) and the other 30% is degassed by life (biological pump).[338] If these biological processes were removed, the level of atmospheric CO_2 would increase five fold.[339] There is a perception that with increased emissions of CO_2 into the atmosphere, the atmospheric CO_2 content will increase forever. This is not the case as atmospheric CO_2 is constantly recycled through the oceans. Any slight variation in organisms in the oceans could account for variations far greater than the human input of CO_2 into the atmosphere.

At a depth of 10 metres, seawater can dissolve twice its own volume of CO_2. The amount of CO_2 that can dissolve in seawater increases with decreasing temperature and increasing pressure. Cold more saline seawater at a high pressure at the bottom of the oceans contains a huge amount of CO_2. When this water rises to the surface, CO_2 is released. We see the same process with aerated drinks. When the pressure is released (i.e. opening the bottle or can), CO_2 bubbles appear and bubbles continue to rise as the drink warms to room temperature.

The exchange of CO_2 between the atmosphere and ocean is well known.[340,341] This gives us an upper limit on how much CO_2 con-

338: Volk, T. and Liu, Z. 1988: Controls of CO_2 sources and sinks in the Earth scale surface ocean: temperature and nutrients. *Global Biogeochem. Cycles* 2: 73-89
339: Eriksson, E. 1963: Possible fluctuations in atmospheric carbon dioxide due to changes in the properties of the sea. *Jour. Geophys. Res.* 68: 3871-3876
340: Revelle, R. and Suess, H. E. 1957: Carbon dioxide exchange between atmosphere and ocean and the question of an increase of atmospheric CO_2 during the past decades. *Tellus* 9: 18-27
341: Skirrow, G. 1975: The dissolved gases – carbon dioxide. In: *Chem. Ocean. Vol. 2, 2nd Ed.* (Eds Riley, J. P. and Skirrow, G.), *Academic Press*, 192p

centration in the atmosphere will rise if all available fossil fuel is burned. In order to permanently double the current level of CO_2 in the atmosphere and keep the oceans and atmosphere balanced, the atmosphere needs to be supplied with 51 times the present amount of atmospheric CO_2. The total amount of carbon in known fossil fuel resources could only produce 11 times the amount of CO_2 in the atmosphere.[342] Unless we change the fundamental laws of chemistry and change the way in which oceans work, humans do not have enough fossil fuel on Earth to permanently double the amount of CO_2 in the atmosphere.

If humans burned all the available fossil fuels over the next 300 years, there would be up to fifteen turnovers of CO_2 between the oceans and atmosphere and all the additional CO_2 would be consumed by ocean life and precipitated as calcium carbonate in sea floor sediments.[343] Chemistry tells us this. Geology does also because it has happened before when the planet had a far higher atmospheric CO_2 content. This is the coherence criterion of science.

There is a very small addition of CO_2 to the oceans from the burning of fossil fuels. Fossil fuel contains no C^{14} (derived from cosmic radiation and nuclear bombs) and hence the increase in the C^{13} and C^{12} of seawater can be used to calculate the addition of CO_2 derived from coal and oil burning.[344] This calculation ignores the contribution of CO_2 from other sources such as soil bacteria, volcanoes, floating microorganisms and burning of wood, grass, stubble and dung. Even if it is only the surface of the ocean that contains CO_2 of fossil fuel origin and then it is about 3% of the CO_2 in the surface water.[345]

342: Jaworowski, Z. *et al.* 1992: Atmospheric CO_2 and global warming. A critical view, Second Revised Edition. *Norsk Polarinstitutt Meddelelser* 119: 1-76
343: Abelson, P. H. 1990: Uncertainties about global warming. *Science* 247: 1529
344: Key, R. 2006: The dangers of ocean acidification. *Scientific American* March 2006, 58-65
345: Jaworowski, Z. *et al.* 1992: Atmospheric CO_2 and global warming. A critical view, Second Revised Edition. *Norsk Polarinstitutt Meddelelser* 119: 1-76

Volcanoes lurking beneath the waves

Most of the planet's volcanoes are submarine and account for 75% of the heat transferred to the surface from molten rocks.[346] The material for these volcanoes, including CO_2, derives from the mantle of the Earth. Molten rock can have a very high content of dissolved gases such as H_2O and CO_2.[347,348] There are more than three million basaltic submarine volcanoes and only 1,800 terrestrial volcanoes, most of which are andesitic. Experimental studies on the solution of CO_2 in molten rocks shows that far more CO_2 dissolves in basalt than in andesite, the most common rock type in terrestrial volcanoes.

When molten rocks rise and cool, they release monstrous amounts of CO_2 and other gases[349] and this gas release provided the initial high CO_2 content of the Earth's atmosphere before plant life was abundant.[350] Most gas is released before an eruption. After an eruption, volcanoes leak gases for a very long time. Terrestrial volcanoes explode because they suddenly release gas whereas the weight of at least 3 km of water stops submarine volcanoes exploding. In gas rich volcanoes and submarine volcanoes, the most abundant gas is CO_2.

In submarine volcanoes, the CO_2 does not bubble up from the mid ocean ridges and enter the atmosphere. Some submarine volca-

346: Crisp, J. A. 1984: Rates of magma emplacement and volcanic output. *Jour. Volcan. Geotherm. Res.* 20: 177-211

347: Stolper, E. and Holloway, J. R. 1988: Experimental determination of the solubility of carbon dioxide in molten basalt at low pressure. *Earth Planet. Sci. Lett.* 87: 397-408

348: Shilobreyeva, S. N. and Kadik, A. A. 1990: Solubility of CO_2 in magmatic melts at high temperatures and pressures. *Geochem. Internat.* 27: 31-41

349: Marty, B. and Zimmermann, L. 1999: Volatiles (He, C, N, Ar) in mid-ocean ridge basalts: assessment of shallow-level fractionation and characterisation of source composition. *Geochim. Cosmochim. Acta* 63: 3619-3633

350: Bottinga, Y. and Javoy, M. 1989: MORB degassing: evolution of CO_2. *Earth Planet. Sci. Lett.* 95: 215-225

noes and associated springs have pools of liquid CO_2.[351] The high-pressure cool bottom waters dissolve all the volcanic CO_2 and this abundant source of CO_2 never enters the calculations of climate "scientists". Many mid ocean ridge lavas are supersaturated in CO_2[352] and release CO_2 into bottom waters as a part of the normal sea floor spreading process.[353,354,355] The total length of mid ocean ridges in the oceans is about 64,000 km and the amount of CO_2 released is huge.[356] Beneath these mid ocean ridges are huge volumes of molten rock and an accumulation of gas, mainly CO_2.[357]

Volcanic gases such as CO_2 escape from the molten rock prior to eruptions, during eruptions and after eruptions.[358] Unless measurements can be made before, during and after a submarine volcanic eruption with instruments awaiting an eruption, then the amount of CO_2 released into seawater cannot be directly measured. However, if submarine volcanic rocks suddenly freeze to a glass, then they trap

351: http://news.nationalgeographic.com.news/2006/08/060830-carbon-lakes.html

352: Jendrzejewski, N. *et al.* 1997: Carbon solubility in mid-ocean ridge basaltic melt at low pressures (250-1950 bar). *Chem. Geol.* 138: 81-92

353: Gerlach, T. M. 1989: Degassing of carbon dioxide from basaltic magma at spreading centers, II. Mid-ocean ridge basalts. *Jour. Volcan. Geoth. Res.* 39: 221-232

354: Jendrzejewski, N. *et al.* 1992: Water and carbon contents and isotopic compositions in Indian Ocean MORB. *EOS* 73: 352

355: Dixon, J. E. and Stolper, E. M. 1995: An experimental study of water and carbon dioxide solubilities in mid-ocean ridge basaltic liquids, Part II. Applications to degassing. *Jour. Petrol.* 36: 1633-1646

356: Pineau, F. and Javoy, M. 1994: Strong degassing at ridge crests: the behaviour of dissolved carbon and water in basaltic glasses at 14°N (M.A.R.). *Earth Planet. Sci. Lett.* 123: 179-198

357: Hauri, E. *et al.* 1993: Evidence for hot-spot related carbonatite metasomatism in the oceanic upper mantle. *Nature* 365: 221-227

358: Kingsley, R. H. and Schilling, J.-G. 1995: Carbon in mid-Atlantic ridge basalt glasses from 28°N to 63°N: evidence for a carbon-enriched Azores mantle plume. *Earth Planet. Sci. Lett.* 129: 31-53.

CO_2 in the glass.[359,360] Although the molten rock may have lost most of its CO_2 during its rise and cooling, the trapped CO_2 in glass can give a minimum figure for the amount of CO_2 released into bottom ocean water[361].

Not only are large amounts of CO_2 released from rising and cooling molten rocks at or beneath the sea floor at the mid ocean ridges, the hot springs in the mid ocean ridges also release small amounts of CO_2.[362] At the sites of hot springs, there is a slight and local increase in acidity. Submarine hydrothermal hot springs are acid. This is caused by the venting of both CO_2 and sulphuric acid.[363] Submarine hot springs have a highly variable CO_2 content[364] and CO_2 can be removed from hot springs by precipitation of carbonates in fluid-rock reactions. Mid ocean ridge hot springs release an estimated 0.3 to 1.2% of the annual input of CO_2 into the oceans[365] whereas mid ocean ridge volcanoes release a far larger and unknown amount of CO_2 into ocean water via gas vents. Volcanoes add far more CO_2 to the oceans and atmosphere than humans.

359: Des Marais, D. J. and Moore, J. G. 1984: Carbon and its isotopes in mid-oceanic basaltic glasses. *Earth Planet. Sci. Lett.* 69: 43-57

360: Dixon, J. E. and Stolper, E. M. 1995: An experimental study of water and carbon dioxide solubilities in mid-ocean ridge basaltic liquids. Part II: Applications to degassing. *Jour. Petrol.* 36: 1633-1646

361: Marty, B. and Tolstikhin, I. N. 1998: CO_2 fluxes from mid-ocean ridges, arcs, and plumes. *Chem. Geol.* 145: 233-248

362: Resing, J. A., *et al.* 2004: CO_2 and ^3He in hydrothermal plumes: implications for mid-ocean ridge CO_2 flux. *Earth Planet. Sci. Lett.* 226: 449-464

363: Mottl, M. J. and McConachy, T. F. 1990: Chemical processes in buoyant hydrothermal plumes on the East Pacific Rise near 21°N. *Geochim. Cosmochim. Acta* 54: 1911-1927

364: Sansone, F. J. *et al.* 1998: CO_2-depleted fluids from mid-ocean ridge-flank hydrothermal springs. *Geochim. Cosmochim. Acta* 62: 2247-2252

365: LeQuéré, C. and Metzel, N. 2004: Chapter 12: Natural processes regulation the ocean uptake of CO_2. In: *SCOPE 62, The global carbon cycle: Integrating humans, climate, and the natural world* (Eds Field, C. B. and Raupach, M. R.). Island Press, 243-256

Volcanoes in subduction areas where a slab of the crust is being pushed under another crustal slab, also release large amounts of CO_2 from gas vents and hot springs.[366] This process of recycling of the Earth's rocks has been happening for at least 2,500 million years and has been providing variable and large amounts of CO_2 to the Earth's atmosphere during this time.[367] This type of volcano accounts for far less than 1% of all the volcanoes on Earth and yet are the database used by the IPCC to construct volcanic input of CO_2 for their models. These are the terrestrial and island volcanoes we see. These are explosive volcanoes and the explosions are due to the sudden expansion of H_2O and CO_2.

When one slab of crust is pushed underneath another, not only do we release CO_2 from molten rocks but CO_2 is also released during the heating of limey rocks.[368] The volume of CO_2 derived from degassing limey rocks is hundreds of times greater than that released from molten rocks. The sudden release of this CO_2 can form craters 1000 metres in diameter.[369] In the Arctic Ocean, huge deep-water explosive submarine volcanoes have formed large craters that released massive amounts of heat and CO_2 into Arctic waters.[370]

In other areas of the Arctic Ocean, there is submarine volcanic and hot spring activity.[371] This has been ignored by the IPCC and is

366: Hilton, D. R. *et al.* 2006: Controls on the He-C systematics of the Izu-Bon-in-Marianas (IBM) subduction zone. *Geochim. Cosmochim. Acta* 70: doi:10.1016/j.gca.2006.06.507

367: Yamamoto, J. *et al.* 2001: Helium and carbon isotopes in fluorites: implications for mantle carbon contribution in an ancient subduction zone. *Jour. Volcan. Geotherm. Res.* 107: 19-26.

368: Schuiling, R. D. 2005: *Our bubbling Earth.* Elsevier

369: Fytikas, M. 1989: Updating of the geological and geothermal research on Milos Island. *Geothermics* 18: 485-496

370: Sohn, R. A. *et al.* 2008: Explosive volcanism on the ultraslow-spreading Gakkal ridge, Arctic Ocean. *Nature* 453: 1236-1238

371: Snow, J. *et al.* 2001: Magmatic and hydrothermal activity in the Lena Trough, Arctic Ocean. *Trans. Amer. Geophys. Union* 82: 193, 197-198

not used in any climate models. This release of hot CO_2 can heat a large volume of rocks[372] and CO_2 from the mantle can pool beneath the crust for later leakage into the atmosphere.[373] We know that hot springs and gas vents emit CO_2 into the ocean and the atmosphere with most of the gas exhaled from gas vents.

The Amazon River has a plume of low salinity water that stretches 3,000 km into the tropical Atlantic Ocean. One would expect that this body of water mixing with seawater would result in the emission of CO_2 to the atmosphere. However, recent research shows that significant amounts of CO_2 are adsorbed from the atmosphere into this water. Nitrogen-fixing bacteria, reliant on the nutrients in the Amazon River runoff, change the ocean-air balance so that instead of emitting CO_2, the ocean adsorbs it.[374] Carbon sequestration of 15 Mt pa occurs in a region that was thought to emit CO_2 into the atmosphere. Nature continues to amaze us with little surprises and hence science is never settled.

Floating organisms use dissolved CO_2 in ocean water and sunlight for photosynthesis.[375] Not only is CO_2 plant food, it increases the cell size.[376] It is used by organisms to make carbonate shells and, when the floating organism dies, the shell sinks. In this way CO_2 is constantly removed from the oceans, estuaries and lakes. An increase in CO_2 and nutrients in seawater are the main factors that

---

372: Schuiling, R. D. 2004: Thermal effects of massive CO_2 emissions associated with subducted volcanism. *Comptes Rendus Geosci.* 336: 1053-1059

373: Schuiling, R. D. and Kreulen, R. 1979: Are thermal domes heated by CO_2-rich fluids from the mantle? *Earth Planet. Sci. Lett.* 43: 298-302

374: Cooley, S. R. *et al.* 2007: Seasonal variations in the Amazon plume-related atmospheric carbon sink. *Global Biogeochem. Cycles* doi:10.1029/2006Gb002831

375: Arrigo, K. R. *et al.* 1999: Phytoplankton community structure and the drawdown of nutrients and CO_2 in the Southern Ocean. *Science* 283: 365-367

376: Burkhardt, S., *et al.* 1999: Effect of growth rate, CO_2 concentration, and cell size on the stable carbon isotope fractionation in marine phytoplankton. *Geochim. Cosmochim. Acta* 63: 3729-3741

increase the growth rate of the floating organisms. Changes in sea surface temperature change the amount of heavy and light oxygen in the shells[377] and life preferentially accumulates the light form of carbon.[378,379] Fossilised floating organisms can be used to track sea surface temperature changes as a proxy for climate change and can be used to calculate the CO_2 content of atmospheres in former times.

The cold Southern Ocean sucks huge amounts of CO_2 from the atmosphere. Climate "scientists" thought that this process had stalled due to human activity. Further work showed that the amount of CO_2 being sucked up by the Southern Ocean is actually rising.[380] The more scientific research that is undertaken on the widely held certainties in climate "science", the more complex and less understood they become. Planet Earth is complicated and full of surprises.

Anaemic oceans

Most of the world's oceans are anaemic. They are extremely depleted in iron because they are not acid and contain dissolved oxygen. Iron is a micronutrient for photosynthetic microorganisms. During ice ages, forests and grassland turn to desert and wind increases. The increased wind blows red iron-bearing desert dust into the oceans. This results in a blooming of microorganisms in the oceans.

377: Wolf-Gladrow, D. A., *et al.* 2002: Direct effects of CO_2 concentration on growth and isotopic composition of marine plankton. *Tellus* 51: 461-476

378: Peterson, B. J. and Fry, B. 1987: Stable isotopes in ecosystems studies. *Ann. Rev. Ecol. System.* 18: 293-320

379: Descolas-Gros, C. and Fontungne, M. 2006: Stable carbon isotope fractionation by marine phytoplankton during photosynthesis. *Plant Cell Envir.* 13: 207-218

380: Landschützer, P. *et al.* 2015: The reinvigoration of the Southern Ocean carbon sink. *Science*, doi:10.1126/science.aab2620

The blossoming of these photosynthetic organisms withdraws even more CO_2 from the atmosphere.[381] If CO_2 drives climate, then during an ice age red dust blown into the oceans would accelerate the removal of CO_2 from the atmosphere and the Earth could then not escape from a runaway ice age. This has not happened. Clearly CO_2 is not the main driver of climate.

Ocean pH

The term pH, meaning *pondus Hydronium* or the power of hydrogen, is a numerical measure of the range from extreme acidity to extreme alkalinity. Oceans have a pH of 7.9 to 8.2. This figure is larger than neutral (pH = 7) which means that the oceans are alkaline. The pH scale ranges from 0 to 14, pH 6 is ten times more acid than pH 7 and pH 5 is a hundred times more acid than pH 7. The pH scale is not linear, it is logarithmic hence to acidify seawater from pH 8 to pH 6, an extraordinarily large amount of acid is needed. If acid is present, sediments, rocks and shells become very reactive. These reactions neutralise acid and the oceans return to their normal alkaline state.

The most alkaline waters in the oceans occur in the centre of ocean circulation patterns whereas less alkaline waters occur at sites of upwelling where deep ocean water is brought to the surface. Deep ocean water brought to the surface at these upwelling sites has a higher CO_2 content. As a result photosynthetic microorganisms thrive and become the bottom of the food chain for other abundant marine life. It is no surprise that the great fishing fields of the world occur at sites of upwelling.

381: de Baar, H. J. W. *et al.* 1999: Importance of iron for plankton blooms and carbon dioxide drawdown in the Southern Ocean. *Nature* 373: 412-415

Time and pH

If CO_2 dissolves in seawater, the oceans should become more acid,[382] so the theory goes. If the oceans become acid (pH <7), then it is argued shells of marine organisms could dissolve. This is greatly exaggerated in the popular press as a potential environmental catastrophe in the popular press.[383] The geological record does not show that shells dissolve otherwise there would be no shelly fossils. The oceans are saturated with calcium carbonate to a depth of 4.8 km. This means that if any more CO_2 were added to the oceans, then calcium carbonate would precipitate.[384] Solid calcium carbonate contains 44% by weight of the gas CO_2. However, buffering of seawater prevents oceans becoming acid.

If there were only a small amount of CO_2 dissolved in water, calcium sulphate known as gypsum[385] precipitates instead of calcium carbonate. This has not been found in the ocean but occurs in terrestrial warm fresh water lakes. If the atmospheric CO_2 content was extremely high, the calcium magnesium carbonate dolomite would precipitate from the oceans.[386] This can be shown experimentally, thermodynamically and from geology. This is yet another example of the coherence criterion in science.

382: $CO_2 + H_2O \Leftrightarrow H_2CO_3$; $H_2CO_3 \Leftrightarrow H^+ + HCO_3^-$; $HCO_3^- \Leftrightarrow 2H^+ + CO_3^{2-}$. In the oceans at pH 7.9 to 8.2, CO_2 exists as dissolved gas (1%), HCO_3^- (93%) and CO_3^{2-} (8%). Calcium in seawater binds CO_2 into insoluble carbonates of calcium in shells, coral reefs and mineral precipitates ($Ca^{2+}_{[aq]} + CO_3^{2-}_{[aq]} \Leftrightarrow CaCO_3 \downarrow$). Furthermore, trapped seawater in sediments precipitates carbonate cement. By these processes CO_2 is removed from the atmosphere and stored in marine sediments as fossils, cement and rock. $CaCO_3$ plankton shells can dissolve back into seawater at a depth of >4.8 km.

383: Doney, S. C. 2006: The dangers of ocean acidification. *Scientific American*: March 2006, 58-65

384: Broeckner, W. S. *et al.* 1979: The fate of fossil fuel carbon dioxide and the global carbon budget. *Science* 206: 409-418

385: $CaSO_2.2H_2O$

386: $CaMg(CO_3)_2$

During the Precambrian (older than 542 Ma), the atmospheric CO_2 content was >1% and huge volumes of dolomite were precipitated thereby extracting CO_2 from the oceans and ultimately the atmosphere. Solid dolomite contains 48% CO_2 gas by weight. This means that the balance of CO_2 between the oceans and atmosphere we see today has not changed for thousands of millions of years.[387] This balance was not changed during times of intense sudden release of CO_2 from volcanoes. There is a correlation between increased volcanic production of CO_2 and increased sedimentation of calcium carbonate from the oceans.[388] Again, we see that planet Earth has been remarkably stable for billions of years. The geological processes of carbonate precipitation in the oceans that have taken place for billions of years are ignored in the computer climate models of the IPCC.

Methane

During past times of high atmospheric CO_2, plant growth has accelerated. In warmer climates where microorganisms are very active, plant material decomposes to methane and the atmospheric methane content rises slightly. The main coal formations were deposited when there was a high atmospheric methane and CO_2, plant decomposition was slow during cold times or at high latitudes and hence there could be an accumulation of plant material that would not rot quickly.[389,390]

During the Carboniferous to Permian glaciation from 360 to 260 million years ago, the high atmospheric methane and CO_2 clearly did not produce global warming. This is just one of the numerous nails

---

387: Holland, H. D. 1984: *The chemical evolution of the atmosphere and oceans*. Princeton University Press
388: Budyko, M. I. *et al.* 1987: *History of the Earth's atmosphere*. Springer
389: Diessel, C. F. K. 1992: *Coal-bearing depositional systems*. Springer
390: Bartdorff, O. *et al.* 2008: Phanerozoic evolution of atmospheric methane. *Global Biogeochem. Cycles* 22: GB1008, doi: 10.1029/2007BG002985

in the coffin for the idea that a high atmospheric CO_2 content drives global warming. The decomposition of biological material forms methane that quickly oxidises in the atmosphere to CO_2 and H_2O. In cool marine conditions, methane gas in sediments bonds onto water to form methane hydrate. The sudden release of methane hydrate into the atmosphere can occur with earth tremors, submarine volcanic activity, meteorite and comet impact and submarine debris flows whereas the gradual release can occur with unloading of sediment with sea level fall and warming of sea water.

Methane hydrates in the oceans

A huge future source of methane as an energy source is methane hydrate, frozen water-bearing methane, that occurs in shallow marine settings. In polar areas, methane hydrate occurs in sediments that are at 0°C or less whereas at lower latitudes, it occurs in sediments from 300 to 2,000 metres depth where the bottom water temperature is less than 2°C. What this means is methane hydrates are present on almost all continental shelves and shallow marine basins. Methane hydrate is a huge long-term energy source and the Japanese government agency JOGMEC had already drilled experimental wells for methane hydrate exploitation.

The US Information Administration estimates that methane hydrates contain more energy than all other fossil fuels combined. They could hold between 10,000 and 100,000 trillion cubic feet of gas. By contrast, they estimated that there are 7,000 trillion cubic feet of recoverable shale gas. Some of us have no concerns about "peak oil" or "peak gas". "Peak oil" was calculated on the basis of successful vertical oil wells. Many old and exhausted oil fields are now being drilled with horizontal wells and fracking of these wells has produced huge new reserves. The concept of "peak oil" has disappeared.

Some 55 million years ago, there was a short period (~10,000 years) of rapid warming[391] resulting from a sudden release of major quantities of methane hydrate from the ocean floor. Methane hydrate oxidised in the atmosphere to CO_2 and water vapour and there were acid oceans for a short time.[392,393] Understanding this period of warming and ocean acidity comes from the study of the oxygen and carbon isotopes in calcium carbonate shells.[394] In what was a catastrophic sudden change in warmth and acidity, these shells did not dissolve. They evolved.[395] During the period of sudden warming 55 million years ago, the depth at which calcium carbonate dissolves in the ocean moved from 4.8 to 2.7 km. A very large mass of CO_2 was dissolved in the oceans and this was permanently sequestered by submarine weathering processes over a period of 100,000 years.[396] Methane hydrate is not factored into IPCC models. Why not?

This short warming event 55 million years ago should have produced a sea level rise. Evidence from marine sediments from the New Jersey Shelf, the North Sea and the New Zealand Shelf shows that sea level started to rise 200,000 years before the event and peaked at the warming event. This sea level rise probably resulted from the

391: Nunes, F. and Norris, R. D. 2006: Abrupt reversal of ocean overturning during the Palaeocene/Eocene warm period. *Nature* 439: doi:10.1038/nature04386

392: Pearson, P. N. and Palmer, M. R. 1999: Middle Eocene seawater pH and atmospheric carbon dioxide concentrations. *Science* 284: 1824-1826

393: Zachos, J. C. *et al.* 2005: Rapid acidification of the ocean during the Paleocene-Eocene thermal maximum. *Science* 308: 1611-1615

394: Pak, D. K. and Miller, K. G. 1992: Paleocene to Eocene benthic foraminiferal isotopes and assemblages: implications for deepwater circulation. *Palaeocean.* 7: 405-422

395: Kelly, D. C. *et al.* 1998: Evolutionary consequences of the latest Paleocene thermal maximum from tropical planktonic foraminifera. *Palaeogeogr. Palaeoclimat. Palaeoecol.* 141: 139-161

396: Zachos, J. C. *et al.* 2005: Rapid acidification of the ocean during the Paleocene-Eocene thermal maximum. *Science* 308: 1611-1615

establishment of a submarine super volcano (the North Atlantic Igneous Province[397]), which displaced ocean water. It is clear that warming events and sea level changes are not fully understood and hence cannot be used in computer models. However, many green left environmental activists claim that if humans keep emitting CO_2 from burning fossil fuels, then there will be a repeat of the 55 Ma event. The sudden release of methane, a greenhouse gas 23 times more powerful than CO_2, is a very different thing from a slow release of CO_2 from burning fossil fuels.

Influence of rainwater

Although rainwater is slightly acidic (pH 5.6), by the time it runs over the surface and chemically reacts with minerals in soils and rocks, it enters the oceans as alkaline water.[398,399,400] The salts transported down rivers result from rainwater reacting with rocks making river water alkaline and slightly saline.[401] Soils contain far more CO_2 than the atmosphere. During weathering, soils release a huge amount of CO_2 which ends up in river systems[402] and the total dissolved CO_2 in river systems depends upon the season, the position of the water in

397: Saunders A. D. *et al.* 1997: The North Atlantic Igneous Province. *Amer. Geophy. Union* Monograph 100, 43-91

398: Velbel, M. A. 1993: Temperature dependence of silicate weathering in nature: How strong a negative feedback on long-term accumulation of atmospheric CO_2 and global greenhouse warming? *Geology* 21:1059-1061

399: Kump, L. R. *et al.* 2000: Chemical weathering, atmospheric CO_2 and climate. *Ann. Rev. Earth Planet. Sci.* 28: 611-667

400: Gaillardet, J. *et al.* 1999: Global silicate weathering and CO_2 consumption rates deduced from the chemistry of large rivers. *Chem. Geol.* 159: 3-30

401: Karim, A. and Veizer, J. 2000: Weathering processes in the Indus River Basin: implications from riverine carbon, sulfur, oxygen, and strontium isotopes. *Chem. Geol.* 170: 153-177

402: Telmer, K. and Veizer, J. 1999: Carbon fluxes, pCO_2 and substrate weathering in a large northern river basin, Canada: carbon isotope perspectives. *Chem. Geol.* 159: 61-86

the river system and whether dissolved carbon has been converted to CO_2.[403]

This process, weathering, has been removing CO_2 from the atmosphere and soils for billions of years and storing the CO_2 in rocks.[404,405,406] If an addition of CO_2 to the atmosphere drives global warming as the green left environmental activists tell us, then removal of CO_2 from the atmosphere should trigger glaciation. It doesn't so we must reject the unsubstantiated opinion that atmospheric CO_2 drives climate change. The higher the temperature and CO_2 content, the quicker the removal of CO_2 by calcium carbonate precipitation.[407] Over time, there has been a balance between CO_2 uptake by soils, rocks, water and life and release of CO_2 into the atmosphere.[408] This has resulted in the long-term stabilisation of the global surface tem-

---

403: Barth, J. A. C. and Veizer, J. 1999: Carbon cycle in St. Lawrence aquatic ecosystems at Cornwall (Ontario), Canada: seasonal and spatial variations. *Chem. Geol.* 159: 107-128

404: Berner, R. A. *et al.* 1983: The carbonate-silicate geochemical cycle and its effect on atmospheric carbon dioxide over the past 100 million years. *Amer. Jour. Sci.* 283: 641-683

405: Raymo, M. E. and Ruddiman, W. F. 1992: Tectonic forcing of late Cenozoic climate. *Nature* 359: 117-122

406: $CO_2 + H_2O \Leftrightarrow H_2CO_3$; $H_2CO_3 \Leftrightarrow H^+ + HCO_3^-$; $2Ca^{2+} + 2HCO_3^- + KAl_2AlSi_3O_{10}(OH)_2 + 4H_2O \Leftrightarrow 3Al^{3+} + K^+ + 6SiO_2 + 12H_2O$; $2KAlSi_3O_8 + 2H^+ + H_2O \Leftrightarrow Al_2Si_2O_5(OH)_4 + 2K^+ + 4SiO_2$; $2NaAlSi_3O_8 + 2H^+ + H_2O \Leftrightarrow Al_2Si_2O_5(OH)_4 + 2Na^+ + 4SiO_2$; $CaAl_2Si_2O_8 + 2H^+ + H_2O \Leftrightarrow Al_2Si_2O_5(OH)_4 + Ca^{2+}$; $KAl_2AlSi_3O_{10}(OH)_2 + 3Si(OH)_4 + 10H^+ \Leftrightarrow 3Al^{3+} + K^+ + 6SiO_2 + 12H_2O$; $CO_2 + CaSiO_3 \Leftrightarrow CaCO_3 + SiO_2$; $CO_2 + FeSiO_3 \Leftrightarrow FeCO_3 + SiO_2$; $CO_2 + MgSiO_3 \Leftrightarrow MgCO_3 + SiO_2$

407: Walker, J. C. B. *et al.* 1981: A negative feedback mechanism for the long term stabilization of the Earth's surface temperature. *Jour. Geophys. Res.* 86: 9776-9782

408: Berner, R. A. 1980: Global CO_2 degassing and the carbon cycle: comment on 'Cretaceous ocean crust at DSDP sites 417 and 418: carbon uptake from weathering vs loss by magmatic activity." *Geochim. Cosmochim. Acta* 54: 2889

perature. Even if this geological stabilisation did not occur, there still would be long term stabilisation by life.[409]

Rainwater runs into and accumulates in freshwater lakes, which are invariably slightly acid because of the lack of major buffering and organic material.[410,411] Yet lakes commonly contain shells of floating organisms and macrofauna,[412,413] especially those that are alkaline.[414] If lakes become extremely acid, life dies.[415] Freshwater lakes also have an excess of calcium[416] and some lakes and seas (e.g. Black Sea) have oxygen-poor bottom waters. Apart from a few short sharp events,[417] the oceans have remained alkaline for billions of years. The oceans have removed dissolved CO_2 by precipitation of calcium carbonate minerals in shells, coral reefs, cement that binds mineral grains together and mineral deposits. Because the oceans have an

409: Schwartzman, D. W. and Volk, T. 1989: Biotic enhancement of weathering and the habitability of Earth. *Nature* 311: 45-47

410: Dermott, R. *et al.* 1986: The benthic fauna of 41 acid sensitive headwater lakes in North Central Ontario. *Water Air Soil Poll.* 28: 283-292

411: Harvey, H. H. and McArdle, J. M. 2004: Composition of the benthos in relation to pH in the LaCloche lakes. *Water Air Soil Poll.* 30: 529-536

412: Pip, E. 1987: Species richness of freshwater gastropod communities in central North America. *Malacol. Soc. Lond.* 53: 163-170

413: Rintelen, T. von and Glaubrecht, M. 2003: New discoveries in old lakes: Three new species of *Tylomelania* Sarasin & Sarasin, 1897 (Gastropoda: Cerithioidea: Pachychilidae) from the Malili Lake system on Sulawesi, Indonesia. *Malacol. Soc. Lond.* 69: 3-17

414: Bennike, O. *et al.* 1998: Fauna and flora in submarine early Holocene lake-marl deposits from the southwestern Baltic Sea. *The Holocene* 8: 353-358

415: Nilssen, J. P. 1980: Acidification of a small watershed in southern Norway and some characteristics of acidic aquatic environments. *Internat. Revue der Gesamten Hydrobiologie* 65: 177-207

416: For example, the weathering of limestone by acid rains produces calcium for accumulation in lakes and oceans. $CO_2 + CaCO_3 + H_2O \Rightarrow 2(HCO_3)^- + 2Ca^{2+}$

417: Lowenstein, T. K. and Demicco, R. V. 2006: Elevated Eocene atmospheric CO_2 and its subsequent decline. *Science* 313: 1928

excess of calcium, if more CO_2 is dissolved in the oceans, then more calcium carbonate is precipitated. While the oceans have an excess of calcium, they cannot become acid and will remain in equilibrium. Calcium is continually added to seawater because it is dissolved in river water.

Acid enters the ocean from submarine hot springs such that seawater around hot springs can be less alkaline than elsewhere. This is especially the case with hot springs that are close to the coast in populated areas where water clarity, sediment deposition, shelter from waves, runoff of rainwater and human influences (e.g. sewage) change alkalinity. In these cases, short periods of decreased alkalinity lead to a decrease in animals that graze on green algae and hence an increase in green algae.[418]

Buffers: Why the oceans will not become acid

A reaction between seawater and minerals on the ocean floor keeps the oceans alkaline.[419] This is called buffering. The floor of the oceans is covered with the volcanic rock basalt. This is a highly reactive rock, especially when it is glassy.[420] Seafloor basalts are fractured allowing the ingress of seawater. Well known reactions between seawater and basalt make the oceans more alkaline. This balances the acid added by hot springs and balance the addition of CO_2 to seawater. These reactions can be duplicated in experiments, can be calculated ther-

---

418: Hall-Spencer, J. M. *et al.* 2008: Volcanic carbon dioxide vents show ecosystem effects of ocean acidification. *Nature* 453: doi:10.1038/nature07051

419: For example, the weathering of silicates such as pyroxenes consumes CO_2 and forms carbonates. The same reactions apply for olivines, a far more reactive family of minerals than the pyroxenes. $CO_2 + CaSiO_3 \Leftrightarrow CaCO_3 + SiO_2$; $CO_2 + FeSiO_3 \Leftrightarrow FeCO_3 + SiO_2$; $CO_2 + MgSiO_3 \Leftrightarrow MgCO_3 + SiO_2$

420: Feldspars are the most abundant minerals in terrestrial and submarine rocks and buffer acidity by reaction to form kaolinite. $2KAlSi_3O_8 + 2H^+ + H_2O \Leftrightarrow Al_2Si_2O_5(OH)_4 + 2K^+ + 4SiO_2$; $2NaAlSi_3O_8 + 2H^+ + H_2O \Leftrightarrow Al_2Si_2O_5(OH)_4 + 2Na^+ + 4SiO_2$; $CaAl_2Si_2O_8 + 2H^+ + H_2O \Leftrightarrow Al_2Si_2O_5(OH)_4 + Ca^{2+}$

modynamically and can be observed in oceanic settings. Again, this is the coherence criterion of science. Over time, the basalt-seawater reactions have controlled the atmosphere and seawater chemistry.[421] Seawater in contact with ocean floor rocks, especially basalt, also removes CO_2 from seawater to form carbonates.[422] There is a fine balance in the oceans where microorganisms consume CO_2 as plant food. This makes seawater more alkaline whereas the decomposition of organisms makes the oceans more acid. The more CO_2 in the atmosphere, the more microorganisms thrive in the oceans.

These mineral and biological processes have taken place for billions of years and modern and ancient seafloor basalts shows that the oceans have been stopping an increase of acidity even when the CO_2 in the atmosphere was 25 times the current amount. Despite huge changes in atmospheric CO_2 over the last few hundred million years, average global temperature has not changed by more than $\pm 3.5°C$, oceans have not become acid and there has been no runaway greenhouse.[423] What a remarkable stable planet we live on where everything is at equilibrium. Fossil shells, algal reefs and coral reefs in ancient rocks show that oceans could not possibly have been acid at times when atmospheric CO_2 and temperature were far higher than now. Plants also thrived at those times.[424]

In fact, the higher the atmospheric CO_2 in the past, the easier it has been to form shells. If the oceans were acid, shells would dissolve and the oceans would become alkaline. Furthermore, at these

421: Arvidson, R. S. *et al.* 2005: The control of Phanerozoic atmosphere and seawater composition by basalt-seawater exchange reactions. *Jour. Geochem. Explor.* 88: 412-415

422: $Ca^{2+} + H_2O + CO_2 \Leftrightarrow CaCO_3 + 2H^+; H^+ + (OH)^- \Leftrightarrow H_2O$

423: Royer, D. L., Berner, R. A. and Park, J. 2007: Climate sensitivity constrained by CO_2 concentrations over the past 420 million years. *Nature* 446: 530-532

424: Bice, K. L. *et al.* 2003: Extreme polar warmth during the Cretaceous greenhouse? Paradox of Turonian $\partial^{18}O$ record at Deep Sea Drilling Project Site 511. *Palaeoceanography* 18:1-11

times the high temperature and high CO_2 content of the atmosphere are unrelated.[425,426] Geological history shows us that for the oceans (and land plants) to efficiently fix atmospheric CO_2 and store it in the rocks, the atmospheric CO_2 content needs to be far higher than at present.

Seawater salinity

The salinity of the oceans has been almost constant for billions of years.[427] The earliest oceans may have been slightly warmer and more saline than the modern oceans.[428] Acid rain leaches salts from rocks on the land. These salts are transported by rivers and accumulate in the oceans and are constantly recycled.[429] In the oceans, seawater chemically reacts with basalts and this adds more salts to the water. The same water-rock chemical reactions that have kept the oceans saline have also kept them alkaline. If the oceans were becoming acid, then they should also become less saline. This we don't see.

In some places, seawater is trapped in basins that have been isolated. For example, the Strait of Gibraltar has been closed twice in its history. During this time, the evaporation rate in the Mediterranean Sea exceeded the rate of input of river water. The sea evaporated and left large salt deposits on the sea floor. These salt layers are still there. The last time the Mediterranean evaporated was between 5.96 to

425: Veizer, J. *et al.* 2000: Evidence for decoupling of atmospheric CO_2 and global climate during the Phanerozoic eon. *Nature* 408: 698-701

426: Donnadieu, Y. *et al.* 2006: Cretaceous climate decoupled from CO_2 evolution. *Earth Planet. Sci. Lett.* 248: 426-437

427: Hay, W. W. *et al.* 2001: Evolution of sediment fluxes and ocean salinity. In: *Geologic modeling and simulation: sedimentary systems* (eds Merriam, D. F. and Davis, J. C.), Kluwer, 163-167

428: Knauth, L. P. 2005: Temperature and salinity history of the Precambrian ocean: implications for the course of microbial evolution. *Palaeogeog. Palaeoclim. Palaeoecology* 219: 53-69

429: Rogers, J. J. W. 1996: A history of the continents in the past three billion years. *Jour. Geol.* 104: 91-107

5.33 million years ago. The Mediterranean was later flooded by water pouring in from the Atlantic Ocean through the re-opened Strait.[430]

Changes in seawater over time

Seawater evolves over time. This can be measured by accurate determinations of isotopes[431] in shells of known ages. Although the Earth's atmosphere has been far warmer than at present for most of geological time,[432] changes in shell isotope chemistry may result from an increasing depth of the oceans over the last 500 million years.[433] Seawater evolution can be traced through the evolution of oxygen and strontium isotopes (driven by plate tectonics and the evolution of continents) and the evolution of carbon and sulphur (driven by biological and chemical cycles).

For the last 550 million years, seawater evolution has been unaffected by atmospheric CO_2, despite the gas being more concentrated and temperature being far higher in the past.[434] There is no reason why these tectonic, biogeochemical and geochemical cycles should change just because humans are now on Earth.

Tipping points

A popular catastrophist view is that as the climate warms, less and less CO_2 will be dissolved in the oceans and we will have a "tipping point" when the Earth enters a runway greenhouse climate. Another supposed "tipping point" is that the oceans will become acid. Permanently. There is no such thing as a "tipping point" in science. Use

---

430: Gargarni, J. and Rigollet, C. 2007: Mediterranean sea level variations during the Messinian Salinity Crisis. *Geophys. Res. Lett.* 34: L10405

431: $\partial^{13}C$, $\partial^{18}O$ and $^{87}Sr/^{86}Sr$

432: Apart from the major glaciations

433: Kasting, J. F. *et al.* 2006: Paleoclimates, ocean depth, and the oxygen isotopic composition of seawater. *Earth Planet. Sci. Lett.* 252: 82-93

434: Veizer, J. *et al.* 1999: $^{86}Sr/^{87}Sr$, $\partial^{13}C$ and $\partial^{18}O$ evolution of Phanerozoic seawater. *Chem. Geol.* 161: 59-88

of these words in the popular media and by green left environmental activists immediately advertises non-scientific opinions. These views ignore geological history, natural CO_2 recycling, sequestration of CO_2 by shells and rocks and the logarithmic CO_2-temperature relationship.

Coral reefs are used as the poster child for scares about ocean acidification. Coral reefs build by calcification. Calcium carbonate comprises the hard parts of skeletons. Calcium carbonate contains 44% CO_2. The process of calcification is energetically costly and energy expenditure increases at a lower pH. Marine animals deposit more calcium carbonate when there is more CO_2.[435,436]

In past geological times when there has been far more CO_2 in the atmosphere, the oceans were not acid. There was no tipping point. The oceans are a complex system with internal buffering from boron complexes and the alkaline earth elements and external buffering from the circulation of seawater through the ocean floor sediments and rocks. Many freshwater lakes are acid due to humic material and have a mollusk population with carbonate shells. Bubbling CO_2 into aquariums promotes both plant and coral growth. About 40% of all marine primary production is from diatoms and experiments[437] show that diatoms grow faster under enhanced CO_2.

Experiments on ocean "acidity"

Computer simulations tell a different story and indicate that the

435: Kleypas, J. A. *et al.* 2011: Coral reefs modify their seawater carbon chemistry – case study from a barrier reef (Moorea, French Polynesia). *Global Change Biology* 10.1011/j.1365-2486.2100.02530.x

436: Thomsen, J. *et al.* 2013: Food availability outweighs ocean acidification effects in juvenile Mytilus edulis: laboratory and field experiments. *Global Change Biology* 19: 1017-1027

437: Wu, Y. *et al.* 2014: Ocean acidification enhances the growth rate of larger diatoms. *Limn. Ocean.* 59: 1027-1034

oceans will become acid.[438,439] Experiments with seawater are flawed because they are done in laboratories removed from the ocean floor rocks, sedimentation from continents and flow of river waters into the oceans. It is these processes that have kept the oceans alkaline for billions of years. Laboratory experiments have to provide results in a short time to be reported in scientific journals. Processes over geological time cannot be that easily replicated.

These limited constrained experiments show that when increasing amounts of CO_2 were added to seawater, it became acid and dissolved shells. If a few handfuls of gravel, sediment and clay from the sea floor and some floating photosynthetic life had been added to the experiment to simulate real conditions, then the result would be completely different. Computer simulations that ignore observations and natural processes that have taken place over billions of years end up with a result unrelated to reality. Reality is written in rocks, not models based in incomplete information.

A recent comment[440,441] suggests that decades of experiments testing the effect of elevated atmospheric CO_2 on marine life have failed because of poor experiment design, reporting failures, failure to recognise complexities of ocean chemistry and basic errors in chemistry (especially regarding carbonate solubility). This poor science is driven by the pressure to publish. A few years ago, I had a number of zoologists consult me because they wanted to design experiments to show the effects of increasing atmospheric CO_2 on

438: Caldeira, K. and Wickett, M. 2003: Anthropogenic carbon and ocean pH. *Nature* 425: 365

439: Orr, J. C. *et al.* 2005: Anthropogenic ocean acidification over the twenty-first century and its impact on calcifying organisms. *Nature* 437: 681-686

440: http://www.nature.com/news/crucial-ocean-acidification-models-come-up-short-1.18124

441: Cornwall, C. E. and Hurd, C. L. 2015: Experimental design in ocean acidification research: problems and solutions. *ICES Jour. Mar. Sci.* 118: doi: 10.1093/icesjms/fsv118

marine life. Their experiment was a tank of seawater into which CO_2 was bubbled. They weighed shells immersed in the water during the course of the experiment.

They looked blank when I asked them how the experiment was buffered, how they would replicate river input and how they would replicate the balance between microorganisms and carbonate precipitation. When I asked the researchers why ocean acidification did not happen in the past when atmospheric CO_2 was up to a thousand times higher, blank looks changed to hostile looks. When I asked why dead shells were used rather than live animals with shells that would extract the CO_2 from the seawater during growth, the reply was that this was not what the experiment was about. The whole exercise was to win a research grant, not to undertake science.

Knowledge of the past puts the ocean acidification scare to bed. It is a non-problem touted by the normal rent-seekers. However, the UN is putting out its hand and shrieking that ocean acidification could cost the economy $US1 trillion *per annum* as a result of losses to fishing and tourism. Oyster farm losses were used as an example. However, losses by oyster farmers are normally due to viruses, bacteria or heavy metal contamination. I'm sure that if Western countries provide large amounts of funding for the UN or IPCC then the problem will be solved.

MISINFORMATION

Australia is the largest *per capita* polluter

The statement is often made by green left environmentalists and the uncritical media that Australia is the largest *per capita* polluter in the world. Presumably this is "carbon" pollution (i.e. emissions of CO_2 from domestic and industrial activities) and the use of the words "carbon" pollution by politicians, the media and activists is knowingly deceitful. But then again, deceit is the common thread that unites those promoting human-induced global warming.

The numbers tell a different story. The 23 million people in Australia generate 1.5% of global human-induced emissions of CO_2 annually. Although Australia has 0.33% of the global population, USA emits 14 times and China emits 26 times more CO_2 *per capita* than Australia. Australia's high standard of living, a landmass of 7,692,024 square kilometres with a sparse inland population, the transport of livestock, food and mined products long distances to cities and ports and the export of metals results in high CO_2 emissions from trucks and trains.

Australia exports a significant share of the global refined aluminium, zinc, lead and copper and hence takes a hit for countries that import and use Australia's metals because smelting and refining in Australia results in CO_2 emissions. Annual reported *per capita* CO_2 emissions are in the order of 20 tonnes per person but this is totally misleading. What is not stated is the overwhelming natural sequestration of CO_2 from the atmosphere to vegetation and marine life in Australia. Calculations show that there are 30 hectares of forest and 74 hectares of grassland for every Australian and each hectare sequesters about 1 tonne of CO_2.

Hence on the continental Australian landmass, Australians are removing by natural sequestration more than three times the amount of CO_2 they emit. Australia's net contribution to atmospheric CO_2 is negative and this is confirmed by the net CO_2 flux estimates from the IBUKI satellite CO_2 data set.[442] Is there some sort of medal that the UN can strike for every Australian? Can Australians claim a bundle of money from the UN because they are such fantastic global citizens?

442: http://wattsupwiththat.com/2014/07/05/the-revenge-of-the-climate-reparations/

Australia's continental shelf is 27,450,000 million square kilometres in area. Carbon dioxide dissolves in ocean water and the cooler the water, the more CO_2 dissolves in water. Australia has a huge coastline with currents bringing warm water from the north to the cooler southern waters (East Australian Current, Leeuwin Current) and through the Great Australian Bight (Zeehan Current). Living organisms extract dissolved CO_2 from seawater to build corals and shells. This natural sequestration further locks away Australian emissions of CO_2.

If emissions of CO_2 are a worry to activists, then Australia does more than its fair share to reduce CO_2 from the atmosphere. Do we ever hear this from green left environmental activists? All we hear is constant doomsday stories with activists constantly crying wolf. Furthermore, the main population centres of Australia are in the south east and the forests and grasslands in SE Australia fix about 30% of all of Australia's CO_2 emissions. As per normal, it is rural Australia that carries the population centres of Australia.

Media networks are only too happy to broadcast that Australia is the worst polluter in the world but don't mention that CO_2 is not a pollutant and, if some scientifically misguided activist lawyer makes laws to call CO_2 a pollutant (as the EPA has done in the USA). Australia is therefore adsorbing and fixing CO_2 "pollution" from elsewhere on the planet. However, "worst polluters" is a great headline whereas the facts are boring. For those that espouse that human emissions of CO_2 is the major cause of climate change, then Australia should be viewed as the hero and not the villain.

Facts tend to pour cold water on hysteria.

Uncertainties

Green left environmental activists state with certainty that the global CO_2 content is increasing due to man's activities. Is this really so? There is a simple question that should be asked of every green left environmental activist: Prove to me that human emissions of CO_2 drive climate change? You'll be dead and buried before you get a coherent answer. In case there is an attempt to bamboozle you with "science", you could also ask: "What is the molecular weight of CO_2?" These questions will show whether your green left environmental activist can argue from science or from ideology.

Rigorous scientists are trained as sceptics and must ask questions about the measuring process before the conclusions derived from those measurements can be discussed. As a scientist, the questions I would ask about CO_2 are: When were the measurements done? Who did the measurements? How were the measurements made? What measurements were rejected and did not enter the data set? Why were some measurements rejected and some accepted? Was the data adjusted or modified? How was the average global CO_2 content calculated? Have the measurements been repeated and independently validated? Are the measurements in accord with previous validated measurements?

The International Energy Agency announced that human emissions of CO_2 did not increase from 2013 to 2014.[443] If this is the case, then the suggested steady increase in atmospheric CO_2 must be due to natural sources such as ocean degassing, submarine volcanicity, soil degassing or temperature increase. The surface of the planet has been warming for the last 350 years since the grip of the Little Ice Age (Maunder Minimum), it is no surprise that there would be increased oceanic degassing of CO_2, especially in the tropics. As can be shown from historical,[444] ice core[445] and geological[446] calculations of atmospheric CO_2, there is no causal relationship between

443: www.thegwpf.com/iea-global-co2-emissions-have-stopped-rising/

444: Beck, E.-G. 2008: 50 years of continuous measurement of CO_2 on Mauna Loa. *Energy and Environment* 19: 1017-1028

445: Caillon *et al.* N. 2003: Timing of atmospheric CO_2 and Antarctic temperature changes across Termination III. *Science* 299: 1728-1731

446: Berner, R. A. and Kothavala, Z. 2001: Geocarb III: A revised model of atmospheric CO_2 over Phanerozoic time. *Amer. Jour. Sci.* 301: 182-204

atmospheric CO_2 and climate. For example, a statistical analysis of temperature and atmospheric CO_2 shows that a variation in temperature precedes the corresponding variation in atmospheric CO_2 levels.[447] This is exactly what is expected because CO_2 is less soluble in warmer water and it seems that a significant source of CO_2 emissions are from the waters just south of the equator. Upwelling of deep cold water saturated in CO_2 releases dissolved CO_2 into the atmosphere. This is in accord with other studies.[448]

What is the global average CO_2 content and how is it calculated? These complicated questions are dealt with below.

Direct measurement of atmospheric CO_2

There is a 150-year long record of direct measurement of CO_2 in the atmosphere by the same method.[449] More than 90,000 measurements were made between 1812 and 1961 with an accuracy of 1 to 3%. That is the same accuracy as modern satellite measurements[450] of atmospheric CO_2. There were peaks in atmospheric CO_2 in 1825, 1857 and 1942. In 1942, the atmospheric CO_2 content was higher than now.[451] For much of the 19th century and from 1935 to 1950, the atmospheric CO_2 content was higher than at present and varied considerably.

Yet we are presented with a smooth curve (Keeling Curve) showing a regular increase in atmospheric CO_2 over time. This rings alarm bells to anyone who has undertaken scientific measurements as vari-

447: Humlum, O. *et al*. 2013: The phase relation between atmospheric carbon dioxide and global temperature. *Glob. Planet. Change* 100: 51-69

448: Gaudry, A. *et al*. 1987: The 1982-1983 El Niño: a 6 billion ton CO_2 release. *Tellus* 39B: 209-213

449: Pettenkofer method

450: http://www.jcsda.noaa.gov/documents/seminardocs/CrispOCO20080319.pdf

451: Beck, E. 2007: 180 years of atmospheric CO_2 gas analysis by chemical methods. *Energy and Environment* 18: 259-282

ability is the norm. This curve is an amalgamation of atmospheric CO_2 measurements since 1959 by one method and ice core measurements by a different method. The 90,000 pre-1959 measurements were ignored. Maybe there are good reasons to suspect that the techniques to produce numbers may well be inaccurate or meaningless?

In 1959, the measurement technique was changed to infra-red spectroscopy with the establishment of the Mauna Loa (Hawaii) measuring station by C. D. Keeling. Of interest is that C. D. Keeling wrote that his aim was to report the rise in atmospheric CO_2 from fossil fuel combustion.[452,453] It might have been scientifically more honest to measure the atmospheric CO_2 content out of curiosity rather than to confirm a pre-determined conclusion. C. D. Keeling's son now operates Mauna Loa and, as Beck[36] notes, he *"owns the global monopoly of calibration of all CO_2 measurements."* He is also a co-author of the IPCC reports, which are underpinned by Mauna Loa and all other readings as the gold standard for global CO_2 levels.

In effect, the IPCC and Keeling are the gatekeepers for all the measurements of CO_2 that underpin the theory that human emissions of CO_2 drive climate change. With so much power vested in so few people, no wonder some of us question the independence and veracity of the primary data, especially when contrary historical data is ignored for no valid reason. As Aldous Huxley stated, *"Facts do not cease to exist because they are ignored."*

This infra-red method was chosen because it was quick and cheap compared to the wet chemical technique used since 1812 yet there was never a calibration of the two methods against each other despite measurements overlapping from 1959-1961. This means that post-1961 data cannot meaningfully be compared to pre 1961 data.

452: Keeling, C. D. 1960: The concentration and isotopic abundances of carbon dioxide in the atmosphere. *Tellus* 12: 200-203
453: Keeling, C. D. 1973: Industrial production of carbon dioxide from fossil fuels and limestone. *Tellus* 25: 174-198

However, measurements were calibrated against a gas standard. The raw data from Mauna Loa is "edited" by an operator who deletes what may be considered poor data.

There is no doubt that some data is poor as Hawaiian volcanoes emit CO_2 and updrafts from populated areas bring CO_2 to the mountainous measuring site. Some 82% of the raw data collected is rejected. The remaining 18% is used for statistical analysis.[454,455] With such savage editing of the primary data, I could show any trend that I wanted to show.

The Pettenkofer method measurements in northwestern Europe showed that CO_2 varied between 270 and 380 ppm, with annual means of 315 to 331 ppm. There was no tendency for rising or falling CO_2 levels at any one of the measuring stations over the 120-year period. Furthermore, these measurements were taken in industrial areas during post-World War II reconstruction and increasing atmospheric CO_2 would have been expected. While these measurements were being undertaken in northwestern Europe, a measuring station was established on top of Mauna Loa in order to be far away from CO_2-emitting populated and industrial areas. The volcano Mauna Loa emits large quantities of CO_2, as do other Hawaiian, basaltic and oceanic volcanoes.[456]

During one volcanic eruption, the observatory was evacuated for a few months and there was a gap in the data record that represented the period of no measurement. There are now no gaps in the Mauna

454: Pales, J. C. and Keeling, C. D. 1965: The concentration of atmospheric carbon dioxide in Hawaii. *Jour. Geophys. Res.* 70: 6053-6076.

455: Backastow, R. *et al.* 1985: Seasonal amplitude increase in atmospheric concentration at Mauna Loa, Hawaii, 1959-1982. *Jour. Geophys. Res.* 90: 10529-10540.

456: Ryan, S. 1995: Quiescent outgassing of Mauna Loa Volcano 1958-1994. In: *Mauna Loa revealed: structure, composition, history and hazards* (Eds Rhodes, J. M. and Lockwood, J. P.), American Geophysical Union Monograph 92: 92-115

Loa data set.[457] No wonder some of us are very sceptical about data and the keepers of the data. One wonders how much green left environmental activism has crept into the collectors and keepers of data.

NOAA claims that the global atmospheric CO_2 level has risen from 397.42 ppm in January 2014 to 400.14 ppm a year later.[458] This increase of 2.72 ppm has a standard deviation of 0.57 ppm. However, all is not as it seems. It has long been known that the daily CO_2 content a metre above crops can change by up to 170 ppm with the highest values between midnight and 6 am,[459] This has been known from basic chemistry for a long time.[460]

Which number between <300 CO_2 (6pm) and >450 ppm CO_2 (6 am) is used for compilation of a site average which will then be used to compile a global average? There are mathematical methods of integrating all data over time but these are not used. How is it possible to calculate a global average atmospheric CO_2 content? How is it possible to make a statistically significant estimate of the global annual increase of atmospheric CO_2 over time?

The atmospheric CO_2 measurements at Mauna Loa change daily and seasonally. Both the Mauna Loa (Hawaii) and Point Barrow (Alaska) data show large seasonal variability for their instrumental CO_2 concentrations. Does the global average CO_2 which is measured to hundredths of a part per million consider such variations? Photosynthesis does not take place at night time because there is no CO_2 uptake hence the gas concentration increases and photosynthesis

457: Jaworowski, Z. *et al.* 1992: Atmospheric CO_2 and global warming: a critical review; 2nd Revised Edition. *Norsk Polarinstitutt Meddelelser* 119

458: http://www.esri.noaa.gov/gmd/ccgg/trends/global.html

459: Yu Huning, 1993: Field carbon dioxide flux density. *Jour. Envir. Sciences* 5: 470-480

460: Fergusson, J. E. 1985: *Inorganic chemistry and the Earth, chemical resources, their extraction, use and environmental impact.* Pergamon Press

during sunlight hours results in CO_2 uptake. This changes the atmospheric CO_2 content all the time and hence the time of measurement during the day is important. Is the global average of CO_2 measured for one specific time during the day. We must bear in mind that at any one time around the globe, some measuring stations are in daylight and others in darkness? Traffic and industry also emit CO_2 and these emissions vary during the day.

Down slope winds transport CO_2 from distant volcanoes and increase the CO_2 content. Upslope winds during afternoon hours record lower CO_2 because of photosynthetic depletion of CO_2 in Hawaiian sugar cane fields and forests. The raw data is an average of four samples from hour to hour. In 2004, there were a possible 8,784 measurements. Due to instrumental errors, 1,102 samples have no data, 1,085 were not used because of upslope winds, 655 had large variability within one hour but were used in the official figures and 866 had large hour-by-hour variability and were not used.[461] So what was really being measured and why were 5,076 of the 8,734 measurements acceptable? If this was science that I was undertaking I would be very wary of my own results.

The Mauna Loa CO_2 measurements show variations at sub-annual frequencies associated with variations in carbon sources, carbon sinks and atmospheric transport.[462] Air that arrives during the April–June period favours a lower CO_2 concentration. Seasonal changes derive from Northern Hemisphere deciduous plants that take up CO_2 in spring and summer and release it in autumn and winter due to the decay of dead plant material. Every April, the Northern Hemisphere reduction of atmospheric CO_2 shows that Nature reacts quickly to CO_2 in the atmosphere and how large amounts of CO_2 can be removed from the atmosphere in a very short time. This

461: ftp://ftp.cmdl.noaa.gov/ccg/co2/in-situ
462: Litner, B. R. *et al.* 2006: Seasonal circulation and Mauna Loa CO_2 variability. *Jour. Geophys. Res.* 111, d13104, 10.1029/2005JD006535

is not news. For millennia farmers have called this time the grow-ing season. It also shows that a CO_2 molecule has a short life in the atmosphere.

There may be errors in sampling and analytical procedures.[463] Measuring stations are now located around the world and in isolated coastal or island areas to measure CO_2 in air without contamina-tion from life or industrial activity to establish the background CO_2 content of the atmosphere. The problem with these measurements is that land-derived air blowing across the sea loses about 10 ppm of its CO_2 as the CO_2 dissolves in the oceans. If the ocean is cold, more CO_2 is lost because CO_2 is more soluble in colder water than in warmer water.

A greater problem is that the infra-red absorption spectrum of CO_2 overlaps with that for H_2O vapour, ozone, methane, dinitrogen oxide and CFCs.[464] Some infra-red equipment has a cold trap to re-move water vapour. However, CO_2 dissolves in cold water and some CO_2 is also removed. These other gases are detected and measured as CO_2. Gases such as CFCs, although at parts per billion in the atmosphere, have such a high infra-red absorption that they register as parts per million CO_2. Unless all these other atmospheric gases are measured at the same time as CO_2, then the analyses by infra-red techniques must be treated with great caution.

If the Pettenkofer method was used concurrently with infra-red measurement for validation, then there could be more confidence in the infra-red results. The infra-red CO_2 figures are now at the level recorded by the Pettenkofer method 50 years ago. Do we really have absolute proof that CO_2 has risen over the last 50 years?

463: Jaworowski, Z. *et al.* 1992: Atmospheric CO_2 and global warming: a critical review; 2nd Revised Edition. *Norsk Polarinstitutt Meddelelser* 119
464: Briegleb, B. P. 1992: Longwave band model for thermal radiation in climate studies. *Jour. Geophys. Res.* 97: 11475-11485

The IPCC's Third Assessment Report of 2001 argued that only infra-red CO_2 measurements can be relied upon and prior measurements can be disregarded.[465] The Pettenkofer results are just a little inconvenient so the prudent IPCC just decided to ignore data contrary to the mantra. In the Pope's Encyclical, he places great faith in the IPCC. The atmospheric CO_2 measurements since 1812 do not show a steadily increasing atmospheric CO_2 as shown by the Mauna Loa measurements. The IPCC chose to ignore the 90,000 precise CO_2 measurements compiled despite the fact that there is an overlap in time between the Pettenkofer method and the infra-red method measurements at Mauna Loa. If a large body of validated historical data is to be ignored, then a well reasoned argument needs to be given. There was no explanation. This is not science.

The annual mean CO_2 atmospheric content reported at Mauna Loa for 1959 was 315.93 ppm. This was 15 ppm lower than the 1959 measurements for measuring stations using the Pettenkofer method in northwestern Europe. By picking a low figure, the rise in atmospheric CO_2 content can be exaggerated. Measured CO_2 at Mauna Loa increased steadily to 351.45 ppm in early 1989.[466]

The 1989 value is the same as the European measurements 35 years earlier by the Pettenkofer method, which suggests problems with both the measurement methods and the statistical treatment of data. In fact, when the historical chemical measurements are compared with the spectroscopic measurements of air trapped in ice and modern air, there is no correlation. Furthermore, measurement at

465: IPCC, 2001: Climate Change 2001. *The scientific basis. Contributions of working group 1 to the Third Assessment Report of the Intergovernmental Panel on Climate Change* (eds Houghton, J. et al.), Cambridge University Press

466: Keeling, C. D. *et al.* 1989: A three-dimensional model of atmospheric CO_2 transport based on observed winds. 1: Analysis of observational data. In: Aspects of climate variability in the Pacific and the Western Americas (Ed. Peterson, D. H.) *American Geophysical Union Monograph* 55: 165-236

Mauna Loa is by infra-red analysis[467,468] and some ice core measurements of CO_2 in trapped air were by gas chromatography[469] hence there is a danger in correlation of measurements.

A pre-IPCC paper used carefully selected Pettenkofer method data. Any values more than 10% above or below a baseline of 270 ppm were rejected.[470] The rejected data included a large number of the high values determined by chemical methods. Were they real or did they not fit the story? The lowest figure measured since 1812 (270 ppm) is taken as a pre-industrialisation yardstick. The IPCC want it both ways. They are prepared to use the lowest number determined by the Pettenkofer method as a yardstick yet reject Pettenkofer method measurements since 1812 when the atmospheric CO_2 was far higher than at present.

The Mauna Loa observatory has the longest continuous measurement of atmospheric CO_2 at the one site by infra-red spectroscopy. It is considered to be a precise measurement of middle tropospheric CO_2 because of the minimal influence of human or vegetation effects, the minimal influence of volcanism and because there has been nearly 50 years of measurement by the same technique.[471] The Mauna Loa record shows an increase in mean annual concentration of CO_2 from 315.98 ppm of dry air in 1959, 377.38 ppm in 2004 to

467: Keeling, R. F. et al. 1996: Global and hemispheric sinks deduced from changes in atmospheric O_2 concentration. Nature 381: 218-221

468: Keeling, C. D. and Whorf, T. P. 2005: Atmospheric CO_2 records from sites in the SIO air sampling network. In: Trends: A compendium of data on global change. Carbon Dioxide Analysis Center, Oak Ridge National Laboratory, TN

469: MacFarling Meure, C. et al. 2006: Law Dome CO_2, CH_4 and N_2O ice core records extended to 2000 years BP. Geophys. Res. Letts 33: L14810

470: Callendar, G. S. 1938: The artificial production of carbon dioxide and its influence on temperature. Quart. Jour. Royal Met. Soc. 66: 395-400

471: Keeling, C. D. et al. 1976: Atmospheric carbon dioxide variations at Mauna Loa Observatory, Hawaii. Tellus 28, 538-551

401.39 ppm in July 2015. The biggest single jump of 2.87 ppm was in the El Niño year 1997-1998, in accord with other increases associated with El Niño events.[472]

For example,[473] the 1982-1983 El Niño event released 6 billion tons of CO_2 into the atmosphere This raises an uncomfortable question for the green left environmental advocates: What happens if the suggested rise in atmospheric CO_2 content is due to natural processes? This question just does not go away and cannot be shouted down by ideology. If oceans emit such a huge amount of CO_2 during an El Niño event, will governments work out a way to tax such emissions?

NOAA is aware that there are variations in the CO_2 content of the atmosphere and that there is a gradient from higher concentration at the poles to lower concentration at the equator. In an attempt to acquire precise measurements that can moderately accurately (1-2 ppm) measure this gradient and better quantify the seasonal variation in atmospheric CO_2 in the Northern Hemisphere, a satellite-based orbiting carbon observatory (OCO) collects the near infra-red spectra of CO_2 and O_2 in reflected sunlight and validates measurements against land stations such as Hawaii on regional scales at monthly intervals.[474,475]

Measurements are not made in high latitude areas and NASA claim to be able to resolve the interference of the H_2O and CO_2 spectra. Satellite and aircraft data shows that the global atmospheric

472: Bacastow, R. B. *et al.* 1980: Atmospheric carbon dioxide, the Southern Oscillation, and the weak 1975 El Niño. *Science* 210, 66-68

473: Gaudry, A. *et al.* 1987: The 1982-1983 El Niño: a 6 billion ton CO_2 release. *Tellus* 39B: 209-313

474: http://www.jcsda.noaa.gov/documents/seminardocs/CrispOCO20080319.pdf

475: http://disc.sci.gsfc.nasa.gov/

CO_2 content varies greatly both laterally and vertically. This again raises the question: How does one compute the average global CO_2 content of the atmosphere if there is variation laterally, vertically and over time? The order of accuracy of the space measurements is parts per million whereas the land-based measurements are hundredths of a part per million.

What is even more interesting is that NASA's OCO data shows that increases in atmospheric CO_2 are not largely due to Western industrial activity. It appears that the atmospheric CO_2 measured by the OCO emanates from rice paddies and rain forests in the Third World. It is clear that we don't really understand CO_2 variations in the atmosphere and any dogmatic statement that increases in atmospheric CO_2 are due to human industrial activity are unrelated to data.

The release of more water vapour and CO_2 into the atmosphere may have contributed to the Late 20th Century Warming in addition to warming from increased solar output and variations in the Sun's electromagnetic and gravitational fields. Again, a fundamental problem exists: How can a global average CO_2 content be calculated when there is so much variance over all scales of time? How can a small change be determined with statistical confidence? Exactly the same problem occurs for the determination of average global temperature.[476]

476: Clark, Roy 2011: *The dynamic greenhouse effect and the climate averaging paradox.* Ventura Photonics

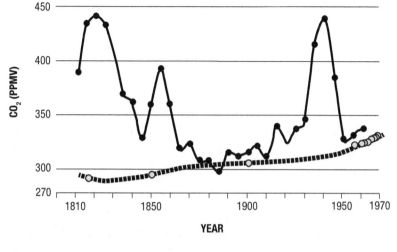

● Mauna Loa 1958   ■■O■■ CO$_2$, 5 year average   ■■■■■ Ice core, Antarctica

Figure 3: Determinations of atmospheric CO$_2$ by the Petten-kofer method (solid line of 5 year averages) between 1812 and 1961, deductions of atmospheric CO$_2$ from Antarctic ice core (gas chromatography) and edited measurements of atmospheric CO$_2$ from Mauna Loa (infra-red spectroscopy, 1958 and onwards).[477] One method of measurement shows great variability in atmospheric CO$_2$ yet another method does not. The high values of CO$_2$ by the Pettenkofer method have been rejected by the IPCC yet the lowest value is used by the IPCC as the baseline pre-industrial value for atmospheric CO$_2$.

Sea surface temperature

Heat in the oceans gives heat to the atmosphere. The heat capacity of water is far higher than the atmosphere and hence the oceans hold the surface heat of the Earth. However, the climate "science" and media propaganda suggests that it is the atmosphere that drives

---

477: Beck, E. 2007: 180 years of atmospheric gas analysis by chemical methods. *Energy Envir.* 18: 259-282

the whole climate system. Because there is a lag between sea surface temperature and atmospheric temperature, the sea surface temperature can be used to try to understand what's just around the climate corner. If we are to use sea surface temperature measurements, then they must be reliable and not "adjusted."

Sea surface temperature used to be measured in the major sea lanes of the world by passing ships. Until recently, sea surface temperature of a bucketful of seawater was measured with a thermometer. These measurements had an accuracy of ±0.5°C rendering such measurements highly inaccurate. However, the change from wooden to canvas buckets led to a "discontinuity" of temperature measurements and significant errors. Currents were also measured by crude methods. It is only in the last 30 years that there have been more accurate and more widespread measurements of sea surface temperature and ocean currents.

Since 1980, floating buoys recorded the temperature of the oceans at a depth of 50 cm and this has been combined with the inlet temperature of ships' cooling water which extracts considerably cooler sea water from a depth of about 10 metres. The estimation of historical sea surface temperature trends by combining bucket measurements, floating buoys and ship inlet water temperature is meaningless. Yet it is this data that is used to calculate changes in historical sea surface temperature. Current and sea surface temperatures older than 30 years are only crude measurements that cannot be used in climate models. However, they are.

There is great uncertainty about ocean warming which only looks at changes since the 1950s. There are large discrepancies between the measuring methods.[478] Since expandable bathythermograph measurements are the largest proportion of the data set, this bias results

---

478: Expendable bathythermographs (XBT), bottle and CTD data. Most data is XBT and this XBT data is positively biased.

in a significant world ocean warming artifact. If the bias is omitted, this reduces ocean heat content change since the 1950s by a factor of 0.62.[479,480]

There are some 3,000 scientific robots in the oceans that have shown a sea surface temperature decrease. The upper layer of the oceans accumulated heat from 1955 to 2003 and then there was a dramatic loss of the average heat content of the upper oceans from 2003 to 2005.[481,482] This is clearly a wake up call for computer climate models. The North Atlantic waters may be currently cooling because the mean annual heat from the Sun at middle and high northern latitudes during the last 11,000 years has decreased.[483] The levels are now those of the Last Glacial Maximum.

Perhaps the cooling of the northwest Atlantic slope waters is heralding the next inevitable glacial period.[484] Although climate over the last 11,000 years is usually described as warm and stable, there have been significant large (e.g. Younger Dryas) and small variations (e.g. Medieval Warming, Little Ice Age). The cooling of slope waters east of the United States and Canada by 4 to 10°C during the Holocene probably resulted from declining heat from the Sun, increasing convection in the Labrador Sea and the equator ward shifting of the Gulf Stream.

479: Gouretski, V. and Koltermann, K. P. 2007: How much is the ocean really warming? *Geophys. Res. Lett.* 34, L01610, doi: 10.1029/2006GL027834

480: Ocean heat content increase (0-3,000m) between 1957-1966 and 1986-1996 estimates of $12.8\pm >8 \times 10^{22}$ Joules.

481: Lyman, J. M. *et al.* 2006: Recent cooling of the upper ocean. *Geophys. Res. Lett.* 33: L18604, doi:10.1029/2006GL027033

482: This equates to a global radiative imbalance of -1.0 $\pm$ 0.3 watts per square metre.

483: Seager, R. *et al.* 2002: Is the Gulf Stream responsible for Europe's mild winters? *Quart. Jour. Royal Met. Soc.* 128: 2563-2586

484: Sachs, J. P. 2007: Cooling of Northwest Atlantic slope waters during the Holocene. *Geophys. Res. Lett.* 34: 10.1029/2006GL028495

Sea surface temperatures can be more reliably measured by using buoys. Both the satellite measurements of lower troposphere temperature and ARGO bathythermograph measurements of sea surface temperature show no warming for at least 11 years. The ARGO system is the best of a bunch of bad sea surface temperature measuring methods. However, it just did not show that the oceans were warming and that we would all fry and die.

The solution was simple: Arbitrarily adjust the sea surface temperature to align ship measurements of sea surface temperature with night marine air temperature estimates despite the fact that they are at odds with all other sea surface temperatures.[485,486] Both data sets have huge problems with reliability. Some 11 unexplained changes to the sea surface data set were made, 0.12°C was added to each buoy to bring measurements in line with those made from ships and changes were made mainly to the highest quality data sets (i.e. post 1998).

Cherry picking the start and finish point of any series of measurements of time can give the required answer, especially if the maximum low or high points are chosen. These sort of statistical games have been played for decades[487] and the climate "scientists" use such techniques all the time. Karl *et al.* start their trend estimates in 1998 and 2000. The year 1998 was an El Niño year. The Karl *et al.* 1998 to 2014 trend (0.106 ±0.058°C per decade) is lower than the 2000 to 2014 trend (0.116 ±0.067°C per decade). This is expected if the trend was started when sea surface temperatures were cooler, as they were in 2000. Again, we see that there is a thread of deceit through most of the climate "scientists'" "science".

The more reliable ARGO buoy data was not used and only the

485: Karl, T. R. *et al.* 2015: Possible artifacts of data bases in the recent global warming surface hiatus. *Scienceexpress*, 4 June 2015

486: http://www.rossmckitrick,com/uploads/4/8/0/4808045/mckitrick_comments_on_karl2015.pdf

487: Darrell Huff 1954: *How to lie with statistics*. Norton

longer--running and more unreliable historical records taken from bucket temperature measurements[488] and boat engine intake cooling water temperature were used. More reliable satellite data was not used.[489,490] Do you think that they were adjusted downwards or upwards? The end result was that the pause in warming over the last 18 years has now been "unfound". It disappeared and that means that there is no pause. Easy. Just change the temperatures upwards by 0.12°C. What does 0.12°C really mean?

The first thing a school pupil learns about science is the order of accuracy. A temperature rise of 0.12°C is, with the order of accuracy, really 0.12±1.7°C[491] hence any conclusions are totally meaningless. The uncritical usual suspects loved it because it satisfied their ideology and even claimed that the highly speculative Karl *et al.* paper should "end the discussion of the so-called pause, which never existed in the first place"[492] and "US scientists: Global warming pause no longer valid".[493] The so-called environmental writers of the global warming cheer squad showed that they know absolutely nothing about measurement in science, know nothing about statistical techniques used to weed out random "noise" and ignore the fact that variations are smaller than the order of accuracy.

Do the experiment yourself. Go to any temperature record. Pick an unusually cold year (say 2000) at the start of your plot and a warm year at the end (say 2014). For God's sake don't even think about

488: Kent, E. C. et al. 2010: Effects of instrumentation changes on sea surface temperature measured *in situ. Climate Change* 1: 718-728

489: http://nsstc.uah.edu/climate/index.html

490: http:///.remss.com/research/climate

491: Kennedy, J. J. *et al.* 2011: Reassessing biases and other uncertainties in sea-surface temperature observations measured in situ since 1850, part 2: biases and homogenisation. *Geophys. Res. Lett.* 116 (D4), doi 10.1029/2010JD015218

492: http://www.theguardian.com/environment/climate-consensus-97-percent/2015/jun/04/new-research-suggests-global-warming-is-accelerating

493: http://www.bbc.com/news/science-environment-33006179

order of accuracy, just plot your graph and claim that the world is warming. Oh … and make sure you use the words "significant trend". It adds gravitas. It's easy. Climate "scientists" do it so why can't you. If you are silly enough to pick the last 2,000 years of temperature from proxies and measurements, then the story is different and is not worth writing about because it's not sensational.

Karl *et al.* (2015) are not alone. The lack of warming for the last 18 years has stimulated more than 30 very creative explanations. For example, it now appears that the variability of climate in the 20th century, on inter-annual to multi-decadal scales, is due to changing ocean circulations.[494] However, no data exists to test this relationship on a longer time scale. It is interesting that Karl who for decades has been promoting CO_2-driven climate change now finds he's in a spot of bother and tells the world that the ocean now drives climate change. As long as the grants keep rolling in I suppose it doesn't matter.

Ice cores

Measurements of CO_2 in ice cores have given further reasons for disquiet. Gas trapped in ice diffuses upwards as a result of the weight of ice. The absolute measurement of CO_2 in ice may therefore not be the CO_2 content of the atmosphere trapped in falling snow. The isotopic C^{12}/C^{13} ratio of CO_2 is temperature dependent if systems are at equilibrium. This is the catch. There is a rapid equilibration between CO_2 in the atmosphere and CO_2 in the oceans.[495]

If ice core data[496] is back calculated to show equilibrium, then the average surface temperature of the ocean was 13°C in 1860 and only

494: Trenberth, K. E. 2015: Has there been a hiatus? *Science* 349: 691-692

495: Rohde, H. 1992: Modeling biogeochemical cycles. In: Butcher, S. S. *et al.* (eds). *Global Geochemical Cycles.* Academic Press, 55-72

496: Hellevang, H. and Aagaard, P. 2015: Kort oppholdstid for karbon I atmosfaren – bevis mot menneskeskapte utslipp. httpgeoforsking.no/ressurser/ klimadebatten/944-kprt-oppholdstid (pers. comm. T. Segelstad)

2°C in 2010.[497] This is obviously wrong. Clearly the ice core data used was not valid, which confirms what we have known for a long time: Ice core data can give unreliable past temperatures and atmospheric CO_2 contents[498] and hence cannot be integrated with modern instrumental measurements to show historical trends of atmospheric CO_2. Yet it is.

The oceans and CO_2

Ocean uptake of CO_2 is from the atmosphere, riverine addition of dissolved CO_2, dissolution of sinking carbonate shells at depths greater than 3.8 km and deep submarine volcanoes. Ocean degassing is a major source of atmospheric CO_2. Cold seawater surface dissolves CO_2; when the seawater is at the tropics it releases CO_2. It is only the surface seawater that is releasing CO_2. Degassing of the oceans is taking place in the tropical waters around Hawaii and is seasonal and interannual (depending upon the El Niño cycles).[499] Deep ocean water is saturated in CO_2, pools of liquid CO_2 lie on the ocean floor, some microbial processes are inhibited and deep ocean water has dissolved huge amounts of CO_2 from submarine volcanoes and springs.[500] There is little mixing between surface and deep ocean waters.

If cold ocean water dissolves atmospheric CO_2 or dissolves volcanic CO_2 deep in the oceans, this CO_2 is not released for thousands of years. If atmospheric CO_2 has increased over the last 150 years, both the lag processes and all processes of release of CO_2 into the

---

497: Pers. comm., carbon isotope geochemist T. V. Segelstad, 20.7.2015

498: Jaworoski, Z. *et al.* 1992: Do glaciers tell a true atmospheric CO_2 story? *Sci. Total Environ.* 114: 227-294

499: Landschützer, P. *et al.* 2014: Recent variability of the global ocean carbon sink. *Global Biogeochem. Cycles* 24: 927-949

500: de Beer, D. *et al.* 2013: Saturated CO2 inhibits microbial processes in CO_2-vented deep-sea sediments. *Biogeosciences* 10: 5639-5649

atmosphere need to be critically evaluated. Cold CO_2-bearing water is carried to the tropics by deep currents. When this water rises in upwelling or is heated in the tropics, it releases CO_2. The release of CO_2 into the atmosphere from the oceans tells us about processes that took place thousands of years ago and provides no information about modern processes.

Increased solar activity warms the surface of the oceans, especially in the tropics. This increases the amount of water vapour and CO_2 in the atmosphere and reduces the uptake of CO_2 by the tropical oceans. As a result, some of the CO_2 that has been attributed to human activity is ultimately of solar origin.[501,502] Furthermore, if the surface of the oceans is enriched in iron, floating organisms are fertilised and there is an accelerated extraction of CO_2 from the atmosphere.[503] Such fertilisation can take place with red iron-rich dust storms bringing desert dust from land to sea.

Ocean processes are very large systems with inputs from the Sun, life, the atmosphere, rivers and continents. To argue, as the IPCC does, that we have an understanding of oceanic processes is an oversimplification.

We know more about the Moon than the oceans.

The life of a CO_2 molecule

If each CO_2 molecule in the atmosphere has a short lifespan there, it means that the CO_2 molecules will be removed fast from the atmosphere to be adsorbed in another reservoir. But how fast? Because atmospheric CO_2 is increasing, it was argued that CO_2 has not been

---

501: Scafetta, N. and West, B. J. 2006a: Phenomenological solar signature in 400 years of reconstructed Northern Hemisphere temperature record. *Geophys. Res, Letts* 33: L17718 doi:10.1029/2006GL027142

502: Scafetta, N. and West, B. L. 2006b: Phenomenological solar contribution to the 1900-2000 global surface warming. *Geophys. Res. Letts* 33: L05708 doi:10.1029/2005GL025539

503: www1.whoi.edu/contacts/steeringcommittee.html

dissolved in the sea and must have an atmospheric lifetime of several hundred years.[504] The IPCC suggests that the lifetime is 50 to 200 years.[505] This has been criticised because the term "lifetime" is not defined[506] and because the IPCC has not factored in numerous known sinks of CO_2.[507,508]

The CO_2 atmospheric lifetime can be calculated by measuring the amount of the carbon isotopes C^{12}, C^{13} and C^{14} in atmospheric CO_2. From this, the CO_2 lifetime can be calculated.[509] This can be double-checked by measuring the amount of the inert but radioactive gas radon (Rn^{222}), the solubility of CO_2 and various complicated carbon isotope calculations. Calculations[510] of the lifetime of atmospheric CO_2 based on natural C^{14} give lifetime values of 3 to 25 years (18 separate studies). Dilution of atmospheric C^{14} from fossil fuel burning gives a lifetime of two to seven years (two separate studies). Atomic bomb C^{14} lifetime gives a value of two to more than 10 years (12 separate studies). Measurements of Rn^{222} give a CO_2 atmospheric

504: Rodhe, H. 1992: Modeling biogeochemical cycles. In: *Global biogeochemical cycles* (eds Butcher, S. S. *et al.*) 55-72, Academic Press

505: Houghton, J. T. *et al.* (Eds) 1990: Climate Change. The IPCC Assessment. Intergovernmental Panel on Climate Change. Cambridge University Press

506: O'Neill, B. C. *et al.* 1994: Reservoir timescales for anthropogenic CO_2 in the atmosphere. *Tellus* 46B: 378-389

507: Jaworowski, Z. *et al.* 1992: Atmospheric CO_2 and global warming: a critical review; 2nd Revised Edition. *Norsk Polarinstitutt Meddelelser* 119

508: Segalstad, T. V. 1996: The distribution of CO_2 between atmosphere, hydrosphere and lithosphere; minimal influence from anthropogenic CO_2 on the global "Greenhouse Effect". *The Global Warming Debate. The Report of the European Science and Environment Forum.* (ed. Emsley, J.), 41-50, Bourne Press

509: Essenhigh, R. H. 2009: Potential dependence of global warming on the residence time (RT) in the atmosphere of anthropogenically sourced carbon dioxide. *Energy and Fuels* 23: 2773-2784

510: Sundquist, E. T. 1985: Geological perspectives on carbon dioxide and the carbon cycle. In: *The carbon cycle and atmospheric CO_2: natural variations Archean to present* (eds Sundquist, E. T. and Broecker, W. S.). American Geophysical Union Monograph 32: 5-59

lifetime of 7.8 to 13.2 years (three separate studies). CO_2 solubility gives an atmospheric lifetime of 5.4 years[511] and C^{12} to C^{13} mass balance value for the lifetime as 5.4 years.[512] There is very little disagreement because the order of magnitude is the same.

The lifetime of CO_2 in the atmosphere is about five years, a number previously acknowledged by the former IPCC chairman Bert Bolin.[513] The short atmospheric lifetime of CO_2 means that about 18% of the atmospheric CO_2 pool is exchanged each year. Those that argue that the measured increase of CO_2 in the atmosphere at Mauna Loa is due to burning of fossil fuels have assumed the lifetime of CO_2 in the atmosphere is 50–200 years.[514] Wrong. Contrary to popular perceptions, if CO_2 is emitted into the atmosphere from human activities, it just does not stay in the atmosphere. It is quickly sequestered for recycling.

There is a considerable difference in the atmospheric CO_2 lifetime between the 37 independent measurements and calculations using six different methods and the IPCC computer model. This discrepancy has not been explained by the IPCC. Why is this important? If the CO_2 atmospheric lifetime was 5 years, then the amount of the total atmospheric CO_2 derived from fossil fuel burning would

511: Murray, J. W. 1992: The oceans. In: *Global Biogeochemical Cycles* (eds Butcher, S. S., Charlson, R. J., Orians, G. H. and Wolfe, G. V.), 175-211, Academic Press
512: Segelstad, T. V. 1992: The amount of non-fossil-fuel CO_2 in the atmosphere. *American Geophysical Union*, Chapman Conference on Climate, Volcanism and Global Change, March 23-27, 1992, Hilo, Hawaii, Abstracts 25
513: Bolin, B. and Eriksson, E. 1959: Changes in the carbon dioxide content in the atmosphere and sea due to fossil fuel combustion. In: *The atmosphere and sea in motion. Scientific contributions to the Rossby Memorial Volume.* The Rockefeller Institute Press: 130-142
514: IPCC, 2001: Climate Change 2001. The scientific basis. Contributions of working group 1 to the Third Assessment Report of the Intergovernmental Panel on Climate Change (eds Houghton, J. T., Jenkins, G. J. and Ephraums, J. J.), Cambridge University Press

be 1.2%,[515] not the 21% assumed by the IPCC. In order to make the measurements of the atmospheric CO_2 lifetime agree with the IPCC assumption, it would be necessary to mix all the CO_2 derived from the world's fossil fuel burning with a different CO_2 reservoir that was five times larger than the atmosphere.[516]

There have been attempts to explain this discrepancy. These involve speculations about unmeasured carbon isotope behaviour[517] and ignore chemical and isotopic experiments that show equilibrium between CO_2 and water within a few hours.[518,519] Furthermore, the proportion of C^{12} and C^{13} measured in atmospheric CO_2 is substantially different from that used in the IPCC model.[520] The IPCC model on how long CO_2 resides in the atmosphere and the amount of fossil fuel CO_2 in the atmosphere is not supported by radioactive or stable carbon isotope evidence and hence the basic assumptions are incorrect.

Maybe the IPCC's troubles derive from concerns that ice cores do not give reliable data for past atmospheres, including the pre-

515: Revelle, R. and Suess, H. 1957: Carbon dioxide exchange between atmosphere and ocean and the question of an increase in atmospheric CO_2 during past decades. *Tellus* 9: 18-27

516: Broecker, W. S. *et al.* 1979: Fate of fossil fuel carbon dioxide and the global carbon balance. *Science* 206: 409-418

517: Oeschger, H. and Siegenthaler, U. 1978: The dynamics of the carbon cycle as revealed by isotope studies. In: *Carbon dioxide, climate and society* (ed. Williams, J.), 45-61, Pergamon Press.

518: Inoue, H. and Sugimura, Y. 1985: Carbon isotopic fractionation during the CO_2 exchange process between air and seawater under equilibrium and kinetic conditions. *Geochim. Cosmochim. Acta* 49: 2453-2460

519: Dreybrodt, W. *et al.* 1996: The kinetics of the reaction $CO_2 + H_2O$ Ù $H^+ + HCO_3^-$ as one of the rate limiting steps in the system H_2O-CO_2-$CaCO_3$. *Geochim. Cosmochim. Acta* 60: 3375-3381

520: Keeling, C. D. *et al.* 1989: A three-dimensional model of atmospheric CO_2 transport based on observed winds: 1. Analysis of observational data. In: *Aspects of climate variability in the Pacific and the Western Americas* (ed. Peterson, D. H.). American Geophysical Union Monograph 55: 165-236

Industrial Revolution atmosphere. Maybe the IPCC's troubles derive from the fact that recent atmospheric CO_2 measurements have been by a non-validated instrumental method[521] where results were visually selected and "edited", deviating from unselected measurements of constant CO_2 levels by the highly accurate wet chemical methods at 19 stations in Europe.[522]

The contribution of fossil fuel emissions of CO_2 to the atmosphere and the lifetime of this fossil fuel-derived CO_2 in the atmosphere, are the cornerstone of the whole IPCC *raison d'être*. There is now no *raison d'être*. For the atmospheric CO_2 budget, marine adsorption and degassing, CO_2 from geological sources (volcanic degassing, metamorphism, mountain building, faulting), CO_2 loss from weathering and CO_2 adsorption by microorganisms must be far more important than assumed by the IPCC. The total of the CO_2 released from fossil fuel burning and biogenic releases (4% atmospheric CO_2) is far less important than the 21% of atmospheric CO_2 assumed by the IPCC.

Geology and the atmosphere

The CO_2 content of the atmosphere is ultimately determined by geological processes. Over the last 4,567 million years, the Earth has degassed about half of its estimated CO_2 by geological processes.[523] Carbon has oxidation states from -4 to +4, so it bonds with more than 80 of the 92 natural elements on Earth. Up to 97% of the planet's carbon is hidden deep in the Earth.[524] Even some meteor-

---

521: Jaworowski, Z. *et al.* 1992: Atmospheric CO_2 and global warming: a critical review; 2nd Revised Edition. *Norsk Polarinstitutt Meddelelser* 119

522: Bischof, W. 1960: Periodical variations of the atmospheric CO_2-content in Scandinavia. *Tellus* 12: 216-226

523: Holland, H. 1984: *The chemical evolution of the atmosphere and oceans.* Princeton University Press

524: Javoy, M. 1997: The major volatile elements of the Earth: Their origin, behaviour and fate. *Geophys. Res. Lett.* 24: 177-180

ites from outer space have a high carbon content (i.e. carbonaceous chondrites). Deep in the Earth are a whole host of carbon-bearing minerals (graphite, diamond, carbonates, carbides, carbon in silicates, carbonaceous compounds), carbon and carbon-bearing gases dissolved in molten rocks, gases (CO_2, CH_4 and higher hydrocarbons) and liquids (CO_2, hydrocarbons, bicarbonate-bearing waters).[525]

It has long been known that the mantle contains carbonates. Melts derived from partial melting of the mantle contain up to 8% dissolved CO_2 at a depth of 125 km and release CO_2 during ascent.[526,527] Some melts are composed only of carbonates. This CO_2 actually fluxes the melting of rocks at high temperature and high pressure. When the molten rock erupts at the surface,[528] it contains only 0.01 to 0.001% CO_2 showing that a huge volume of CO_2 is released before, during and after eruptions. Furthermore, mantle melts are mainly in submarine settings and the CO_2 is dissolved in deep bottom waters and not seen as a stream of bubbles to the surface.

CO_2 is extracted from the mantle by this unseen process which has been taking place for thousands of millions of years. Because ocean floor volcanic rocks of mantle origin have been recycled for thousands of millions of years, it is only possible to calculate the minimum amount of CO_2 degassed over the last 200 million years. There are a few salient facts. It is difficult to put a bucket over a volcano before, during and after eruption to collect the amount of CO_2

525: Hazen, R. M. *et al.* 2012: Carbon in the Earth's interior: Storage, cycling and life. *EOS Trans Amer. Geophys. Union* 93: 17-18

526: Katsura, T. and Ito, E. 1990: Melting and sub-solidus phase relations in the $MgSiO_3$-$MgCO_3$ system at high-pressures – implications to evolution of the Earth's atmosphere. *Earth Phys. Sci. Lett.* 99: 110-117

527: Eggler, D. H. 1978: The effect of CO_2 upon melting in the system Na_2O-CaO-Al_2O_3-MgO-SiO_2-CO_2 to 35 kb, with and analysis of melting in a peridotite-H_2O-CO_2 system. *Amer. Jour. Sci.* 278: 305-343

528: Harris, D. M. and Anderson, A. T. 1983: Concentrations, sources and losses of H_2O, CO_2 and S in Kilauean basalt. *Geochim. Cosmochim. Acta* 47: 1139-1150

released, so we must do the calculations of volcanic CO_2 emissions based on experiments showing how much CO_2 is dissolved in a hot high pressure molten rock at depth and the amount of CO_2 in an erupted lava. The difference is the amount of CO_2 emitted by the Earth for an eruption.

The calculated amount is a minimum because calculations don't consider the amount of CO_2 trapped in pores and fractures, precipitated as carbonates and dissolved in water and steam. A few examples tell the story. When the Icelandic volcano Eldgjá erupted in 934 AD, it released more than 15 times the modern annual amount of CO_2 released globally from burning fossil fuels. No global warming followed this eruption. In the 1783-1784 AD eruptions of Laki (Iceland) over a period of 6 months, 12 times the amount of CO_2 we emit into the atmosphere annually was released. There was no global warming following this eruption. There was actually cooling from the volcanic aerosols in the atmosphere.

We often see drawings of the carbon cycle which only deal with atmospheric and surface carbon. There is another carbon cycle about which we know very little – the deep carbon cycle.[529] The IPCC and green left environmental activists ignore carbon-bearing minerals and fluids deep in the Earth. Maybe they have never pondered the origin of the primordial CO_2 in the atmosphere of the Earth, the moon and other planets. But then again, curiosity is lost when one follows the consensus.

It is geology that kills global warming catastrophism (and creationism) and some groups want this geology hidden from the public. For example, when the IPCC-supporting and climate politics agency CICERO saw information about CO_2 on a Natural History Museum of the University of Oslo site,[530] they demanded that it be removed

---

529: Dasgupta, R. 2103: Ingassing, storage and outgassing of terrestrial carbon through geologic time. *Rev. Mineral. Geochem.* 75: 183-229
530: http://www.nhm.uio.no/nyheter/2010/vulcanutbrudd.html

from the web site. The author, Prof Tom Segelstad, is so well known in the geological community that censorship cannot work and information spreads wide and far. We have known about the Earth's CO_2 contribution to the mantle for decades hence it is hard to censor knowledge that forms part of the body of basic geology.

This degassed CO_2 and methane (CH_4, later oxidised to CO_2 and H_2O) from deep in the Earth is not lost to space, it is stored in rocks (such as limestone, carbon-rich sediments, fossil fuels), in solution in water and life. The balance between degassing CO_2 from the Earth's interior and weathering (i.e. atmosphere-biosphere-hydrosphere-lithosphere system), carbonate sedimentation and biological carbon determines the atmospheric CO_2 content. Seawater alkalinity is controlled by CO_2 uptake during weathering and inorganic and biological carbonate deposition in the oceans are major CO_2 sinks.

These sinks have kept the oceans alkaline for billions of years and will continue to keep the oceans alkaline. When we run out of rocks on Earth, there may be a slim chance that the oceans will acidify. Warmer climates accelerate weathering thereby accelerating the uptake of atmospheric CO_2. The atmospheric CO_2 content is controlled by temperature and it appears unlikely that atmospheric CO_2 drives climate.[531] But we knew this anyway from the 800-year lag of increased CO_2 following temperature rise as measured in ice cores.[532]

The key assertions in the global warming claim is that CO_2 has increased approximately 35% over the last 150 years. This assertion is challenged. The release of CO_2 from cooling molten rocks is unmeasured (e.g. Kamchatka), the release of CO_2 from the greatest biomass on planet Earth (i.e. bacteria in the first four kilometres of the Earth's crust) is unknown, the release of CO_2 from uplift of

531: Kondratyev, K. Y. 1988: *Climate shocks: natural and anthropogenic*. John Wiley
532: Petit, J. R. *et al.* 1999: Climate and atmospheric history of the past 420,000 years from the Vostok ice core, Antarctica. *Nature* 399: 429-426

mountain ranges and alps is unknown and the release of CO_2 from submarine volcanoes is unknown. Furthermore, the amount of degassing from oceanic upwelling bringing water saturated in CO_2 to the surface is unknown.

What does the IPCC say about CO_2? There are many papers that use the same data set as the IPCC and derive a different conclusion. For example, studies in China concluded that global warming is not related only to CO_2 and that the effect of CO_2 on temperature is grossly exaggerated.[533] An assertion by the IPCC is that global atmospheric temperature has risen by 0.7°C since 1850. The bulk of the global temperature rise was before massive industrialisation (1850 to 1940). During the post-World War II economic boom temperature decreased while the industrial emissions of CO_2 greatly increased. From 1976 to 1998 temperature increased and CO_2 emissions increased. Temperature has been static since 1998 during the period when human emissions of CO_2 were the greatest from Asian and Indian industrialisation.

Furthermore, there are uncertainties in thermometer measurements and these are not in accord with balloon and satellite temperature measurements. In addition, temperatures rose naturally since the bitterly cold Maunder Minimum 350 years ago. The imputation is that this temperature rise is due to human activity. The pre-1850 temperature rise cannot be related to industrialisation and hence must be related to natural processes. Maybe this process continues today and is unrelated to CO_2 emissions? The post-1940 increase in atmospheric CO_2 may well be due to the release of CO_2 from the oceans lagging behind the solar-driven heating from 1850 to 1940 or even the Medieval Warming or Roman Warming.

Another assertion of the IPCC is that CO_2 is a greenhouse gas.

533: Zhen-Shan, L. and Xian, S. 2007: Multi-scale analysis of global temperature changes and trend of a drop in temperature in the next 20 years. *Meteor. Atmos. Phys.* 95: 115-121

This assertion does not acknowledge that H_2O is the main greenhouse gas, that CO_2 derived from human activity produces 0.1% of global warming and that there is a maximum threshold concentration level for CO_2, after which an increase in CO_2 has very little effect on atmosphere warming.[534] The current CO_2 content of the atmosphere is the lowest it has been for thousands of millions of years, and life (including human life) has thrived at times when CO_2 has been significantly higher.[535]

The IPCC 2007 report stated that the CO_2 radiative forcing had increased 20% during the last 10 years. Radiative forcing quantifies increases in radiative energy in the atmosphere and hence the temperature. In 1995, there was 360 ppm of CO_2 in the atmosphere, whereas in 2015 it was 400 ppm, some 11% higher. However, each additional molecule of CO_2 in the atmosphere causes a smaller radiant energy increase than its predecessor and the real increase in radiative forcing was 1%. The IPCC have exaggerated the effect of CO_2 20-fold. We are getting used to deceit from the IPCC.

During ice ages such as 140,000 years ago, the CO_2 content of the atmosphere was higher than the pre-industrial revolution chosen figure of 270 ppm.[536] It is clear that CO_2 is not the only factor that controls air temperature, otherwise we could not have ice age conditions with a high atmospheric CO_2 content. Let's have a heretical thought or two. Maybe that great ball of heat in the sky called the Sun has a greater effect on surface temperature and climate change than a trace gas in the atmosphere? Perish the thought. Maybe that dreadfully polluting gas CO_2 is the gas of life? Perish the thought.

534: http://www.geocraft.com/WVFossils/greenhouse_data.html

535: deFreitas, C. R. 2002: Are observed changes in the concentration of carbon dioxide in the atmosphere really dangerous? *Bull. Canad. Petrol. Geol.* 50: 297-327

536: Lorius, C. *et al.* 1990: The ice-core record: climate sensitivity and future greenhouse warming. *Nature* 347: 139-145

The transition from a global ice age to global warming at about 250 million years ago was characterised by huge rises (to 2,000 ppm) and falls (to 280 ppm) in the amount of atmospheric CO_2.[537] During this time plant and animal life thrived. If CO_2 was not recycled and if humans burned all the known fossil fuels on Earth, then the atmospheric CO_2 content would be 2,000 ppm. This would be perfect for plants and animals as they have no problems living with 2,000 ppm and the planet need not necessarily be a hot house.

Over geological time, a correlation between CO_2 and temperature over coarse (10 million-year time scales) and fine (million-year time scales) scale resolutions shows that CO_2 operates together with many factors including solar luminosity, tectonics and palaeogeography. There is a suggestion that the 500 proxy records of CO_2 over the last 500 million years show that all cool events are associated with CO_2 levels below 1,000 ppm but when CO_2 is less than 500 ppm, then this is the recipe for widespread continental glaciation.[538] On a more detailed scale, this generalisation is not true and there is no close correlation.

This is contrary to other studies that show that atmospheric CO_2 in previous cold times was more than 4,000 ppm in the Ordovician-Silurian (450 to 420 million years ago) glaciation and about 2,000 ppm in the Jurassic-Cretaceous (151 to 132 million years ago) glaciation[539]. Whatever the figure, six of the six ice ages on planet Earth were initiated when the atmospheric CO_2 content was higher than at present hence a high atmospheric CO_2 content does not necessarily drive warming. Each ice age had numerous glaciations and intergla-

---

537: Montanez, I. P. *et al.* 2007: CO_2-forced climate and vegetation instability during Late Paleozoic deglaciation. *Science* 315: 87-91

538: Royer, D. L. 2006: CO_2-forced climate thresholds during the Phanerozoic. *Geochim. Cosmochem. Acta* 70: 5665-5675

539: Berner, R. A. and Kothavala, Z. 2001: Geocarb III: A revised model of atmospheric CO_2 over Phanerozoic time. *Amer. Jour. Science* 301: 182-204

cials. However, we are told by the green left environmental activists and the scientifically ignorant activist media that we will fry and die if the atmospheric CO_2 content increases just slightly.

It was argued recently that the current glaciation in Greenland was driven by a low atmospheric CO_2 content.[540] Various trigger mechanisms for the glaciation were tested (orbital, closing the Panama seaway, permanent El Niño, Himalayan uplift) using modelling software.[541] This is the same software that attempts to predict future climates on the basis that CO_2 drives climate. This result is not surprising as the modelling is programmed to produce such a result. What is not discussed is how the atmospheric CO_2 fell and nor is the role of extraterrestrial and solar activity discussed.

The current content of CO_2 in the atmosphere is about 400 ppm. This is commensurate with the modern oscillating climate comprising 90,000 years of glaciation and 10,000 years of interglacial warmth. The variation in CO_2 shows that a climate sensitivity no greater than $\pm 3.5°C$ has probably been the most robust feature of the Earth's climate system over the last 420 million years.[542] Over the long history of planet Earth, the oceans have not boiled, the oceans have not frozen from top to bottom, the ocean volume has been almost constant, the ocean pH and salinity have hardly varied and CO_2 has been in the atmosphere and oceans since the beginning of time. We are indeed fortunate to live on a very stable planet yet the green left environmental activists are telling us that the planet is unstable, the world is about to fall apart and it's all our fault.

As a result of a higher temperature, more CO_2 is outgassed from the oceans. These matters are complicated and equilibrium calcula-

---

540: Lunt, D. J. *et al.* 2008: Late Pliocene Greenland glaciation controlled by a decline in atmospheric CO_2 levels. *Nature* 454: 1102-1105
541: HadCM3
542: Royer, D. L. *et al.* 2007: Climate sensitivity constrained by CO_2 concentrations over the past 420 million years. *Nature* 446: 530-532

tions are difficult because of the large number of unknown factors. For example, at low temperatures such as during glaciation, the stronger winds bring more dust into the ocean basins. This dust contains iron and other elements and this stimulates an increase in biological activity, especially in those parts of the ocean that contain limited nutrients. This biological activity removes CO_2 from the atmosphere, yet there has not been a runaway glaciation. In the scheme of things, CO_2 plays second fiddle in global climate.

Photosynthesis is rapid and there are annual cycles where more CO_2 is consumed in the warm sunny summer months than in winter. There are also lags in the return of CO_2 to the atmosphere if organic material produced by photosynthesis is buried in sediments, peat or limey rocks.

The difference in measurement of CO_2 by various methods is unresolved. The residency of CO_2 in the atmosphere is far less than that used by the IPCC and this greatly affects the estimation of the amount of CO_2 produced by humans. Not all sources and sinks of CO_2 are considered by the IPCC. The calculation of the transfer of CO_2 between the atmosphere and the oceans uses a body of incomplete data. By using various general circulation models, computer models for some strange reason agree with other computer models which used the same data set and methodology.[543] There is clearly a lot more to learn about CO_2.

Uncertainties resolved

And to answer the questions at the beginning of this section:

(i) *When were the CO_2 measurements done?* At different times of the day, at different seasons and at different places knowing full well that atmospheric CO_2 varies verti-

---

543: Le Quere, C. *et al.* 2003: Two decades of ocean CO_2 sink and variability. *Tellus* B 55: 649-656

cally, horizontally and temporally. In essence, the data set is pretty sloppy.

(ii) *Who did the measurements?* A diversity of people with some dispassionately collecting data and others involved in environmental activism and research grant chasing.

(iii) *How were the measurements made?* With at least three different techniques with no cross calibration or concerns about equilibrium.

(iv) *What measurements were rejected and did not enter the data set?* A Hell of a lot.

(v) *Why were some measurements rejected and some accepted?* You tell me.

(vi) *Was the data adjusted or modified?* Calibration of satellite data with ground station measurements involves modification.

(vii) *How was the average global CO_2 content calculated?* With bloody great difficulty. How can someone calculate with statistical significance a rise in the atmospheric CO_2 to fractions of a part per million when variation from night to day can be more than 170 ppm? Furthermore, the final statistics cannot be better than the original data which in itself has errors, is not in accord with data collected by other methods and has a high variability.

(viii) *Have the measurements been repeated and independently validated?* No.

(ix) *Are the measurements in accord with previous validated measurements?* No.

The carbon isotope measurements of atmospheric CO_2 are probably not representative of the burning of fossil fuels, have a contribution from vegetation and are not representative of the global CO_2 in equilibrium with seawater. All we have are non-validated measurements of atmospheric CO_2 that may or may not be correct. These measurements probably don't represent an apparent increase of supply of human emissions of CO_2.[544] Furthermore, no matter what time scale we look at, CO_2 is unrelated to climate change.

Don't accuse CO_2 of anything, it is just innocent plant food! Changes by a few parts per million in atmospheric CO_2 indicate a dynamic planet but are certainly not dangerous.[545] We are constantly bombarded with scare stories about how the global average CO_2 content is increasing. The immediate response is: Good. More plant food can only help all of us. And if a trace gas in the atmosphere increases very slightly, then who cares anyway? And if the atmosphere warms ever so slightly, then who cares?

Where do most people, especially in the Northern Hemisphere, go for their summer holidays? To a warmer climate. We humans evolved in the Rift Valley of Africa in a warm climate and later moved to cooler climates. We humans should be humbled by our insignificance. If the 24-hour clock of the history of the planet started at midnight, then we humans have only been on Earth for the last two seconds.

The computer models of coupled ocean-atmosphere circulation models are often quoted as evidence that the measured atmospheric CO_2 increases are the cause of increased global temperatures. These are not observations. They are numerical theories that are based on

---

544: Segelstad, T. V. 2008: Carbon isotope mass balance modeling of atmospheric vs. oceanic CO_2. *33rd International Geological Congress*, Oslo, Norway, Aug. 2008 www.cprm.gov.br/33IGC/1345952.html

545: deFreitas, C. R. 2002: Are observed changes in the concentration of carbon dioxide in the atmosphere really dangerous? *Bull. Canad. Petrol. Geol.* 50: 297-327

dubious assumptions. The pause in global warming over the last 18 years shows that such models are not related to reality and cannot be used for predictions.

This shows that there is no scientific basis for introducing a carbon tax, emissions trading scheme or any other such euphemism. The pause in global warming may continue, we may see an increase in warming or we may see a decrease in warming. The steady state increase in CO_2 as deduced from measuring stations may cease or even reverse. The best policy decision, after deep thought and an independent analysis of the data, is to do nothing or prepare for the next inevitable glaciation.

The average poor person in Africa is more concerned about the next meal and doesn't give a damn about CO_2. When I read or hear about a climate crisis, I think: What crisis? We've been waiting for a climate crisis for 30 years and measurements of temperature show that there is no crisis. There certainly is another crisis. When models show that temperature is going to rise forever and ever and ever as the diagrams suggest, there is a crisis. Not with the temperature because measurements by balloons and satellites show that there is little variation over the last 35 years. The crisis is that if gullible people, green left environmental activists and the scientifically illiterate media all dogmatically state that the models are correct, then there is a crisis. A crisis in education.

Another crisis is one of science. The models are regarded as correct and the measurements are regarded incorrect. If there is a choice between a model that can be easily tweaked to give the required exaggerated answer or measurements, I know what would give me confidence. It's not that one model exaggerates. An average of 102 IPCC CMIP-5 climate models shows that after 1995, there is no relationship between the models and measurements. The models were telling us what should happen. They didn't. Yet we are still asked to believe in the model-driven human-induced global warming doom that will wipe us out in 100 years time even those models are hopelessly wrong over a 20-year period. This point was made very clearly by Dr John Christy (University of Alabama in Huntsville).[546] If it can't be shown unequivocally that human emissions of CO_2 drive dangerous climate change, then maybe it's time to shelve this crisis and get into a lather about the next alleged crisis.

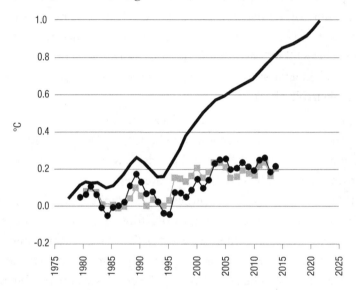

Figure 4: Average of 102 IPCC models (rcp4.5; solid line) and measurements of temperature over the last 30 years (circles are average of 4 balloon data sets, squares are average of 2 satellite data sets)[547]

The Pope needs some perspective on what is only a perceived problem. The amount of industrial CO_2 added to the ocean-atmosphere system (320 Giga tonnes[547]) since the beginning of the Industrial Revolution, which is a tiny fraction of the amount of CO_2 in the

546: http://medi.al.com/news/huntsville/index.ssf/2015/03/climate_expert_john_christy_on.html
547: James Hansen, Congressional testimony

atmosphere (32,000 Giga tonnes[548]). The atmosphere is almost completely opaque to outgoing long-wave radiation and balloon studies have shown that a lower atmosphere convective heat transfer model is in accord with balloon data rather than the IPCC's radiation heat transport model. Very small changes in the global average temperature and atmospheric CO_2 indicate that temperature increases precede atmospheric CO_2 increases with a lag by about 10 months[549].

And why do I labour these points of measurement and CO_2 chemistry? Because the Pope trained in chemistry and worked as a chemist before joining a seminary.[550]

He should have known the uncertainties of measurement and the basics of CO_2 chemistry.

CLIMATE, CATASTROPHISM AND CREATIONISM

Troll trawling

I find it amusing when internet trolls claim that those of us who question human-induced climate change are no different from the dinosaurian modern creationists. One pejorative statement dismisses the need for argument, evidence or even background reading. Well, here is some background reading for trolls. Does any reader want to make a wager whether trolls actually read books?

Creationist and climate catastrophist comparison

For some decades, I have been involved in challenging anti-science, mainly the "creation scientists." This has involved much public activity, litigation and a book.[551] I thank the creationists for giving me a

---

548: IPCC Third Assessment Report

549: Humlum *et al.* 2012: The phase relation between atmospheric carbon dioxide and global temperature. *Glob. Planet. Change* 100: 51-69

550: http://ncronline.org/blogs/ncr-today/does-pope-francis-have-masters-degree-chemistry

551: Plimer, Ian 1994: *Telling lies for God.* Random House

nose for anti-science, scientific fraud and misleading and deceptive conduct. There are many similarities between creationist and green left environmental activists.

Most scientists concentrate on their science. Some are involved in teaching science. A few promote science to the broader community. Science is not conducted by inspecting the work of others as Cook *et al.* did and then deriving breathtaking pre-ordained conclusions. Only a handful of active scientists actually deal with the ingress of anti-science into the schools and community. If litigation, broadcasting, writings and public lectures are any guide, then I am one of those few scientists who tackles anti-science. I am more qualified than any scientist in Australia to make comments on, for instance, creationism.

Creationists, as a matter of faith, believe that planet Earth is a mere 6,000 years old, that all sedimentary rocks and fossils were formed in a great flood some 4,000 years ago and that evolution does not exist. Creationists call their evangelical ministers "creation scientists". Some of these "creation scientists" have basic scientific qualifications and a few have higher degrees. One can count on a saw miller's hand the number who have actually worked as a scientist and published science. Climate activists invented climate "science" and many in the plethora of climate institutes have neither scientific qualifications nor have undertaken science. The "science" of creationists is presented with scriptural authority and intertwined with morality. This is exactly the same for the green left environmental activists who have tried to capture the high moral ground.

Creationists have tried to have their narrow fundamentalist view of the world taught in school science courses and have exerted considerable political pressure to be given a fair go and be allowed to teach an alternative to evolution to every school child in a pluralist society. Because many green left environmental activists are school teachers and university lecturers, they are able to teach their view of the planet to young people unabated.

There is an uncanny similarity between creationists and the climate industry. Just as creationists invented "creation science", the climate industry invented "climate science". Creationist organisations are pretty good businesses. The climate industry is a far better business as government funds seem to be bottomless. Until they're not, as various wind industry groups are finding out in the UK. The business model is the same: Scare people witless, promise eternal salvation and make the gullible pay.

Both creationists and climate catastrophists use a very narrow view of science to come to an all-embracing world-view. Both groups have their prophets (e.g. Duane Gish, Al Gore), a few prominent leaders (e.g. Carl Wieland, Tim Flannery) and an army of the faithful worker bees (e.g. Phil Jones, Michael Mann; Andrew Snelling, Ken Ham). Both groups have a holy book (the Bible; the latest IPCC reports) but very few of the faithful flock have actually read and understood their holy books. The faithful flock just believes. Just ask your average green left environmental activist to show you from first principles how CO_2 drives climate change. They can't. They just believe. Both creationists and climate catastrophists have blind unreasoning faith and entered this position without using critical analysis, logic or knowledge. No amount of critical analysis, logic or knowledge will result in a change of this belief. These beliefs are their reason for living and if such beliefs are removed, then there is no purpose for life.

Both groups publish their own work in their own journals over which they exert strong editorial control. The faithful, be they creation "scientists" or green-left environmental activists, pick over the work of others and don't undertake creative new research themselves. No work that challenges the pre-conceived dogma ever sees the light of day in their journals.

In their various small circles, the officers of both movements profit in terms of money, power and prestige. For some of the green left activists, prestige comes with the number of citations in the literature. For other green left environmental activists, the number of press citations, appearances at inquiries and photo opportunities puts bread on the table.

With both groups, there is a strong underpinning of the principles of Christianity. With the creationists, salvation can be purchased by donations, purchases of literature and DVDs. Both creationists and global warmers have a catastrophist view of the world. They both incorporate aspects of catastrophic scenarios which various unbalanced folk have touted off and on for thousands of years. A negative view of the future predominates. However, all is not lost. Salvation can be purchased (e.g. carbon tax, emissions trading) and, once indulgences have been purchased, then the planet or your soul will be saved.

Both groups are long on cant and short on fact, intertwine data with interpretation and are totally unwilling to expose their primary data to scrutiny. The parallels become even closer when one looks at the use of data. Climate catastrophists and creationists use a very narrow body of data, wallow in trivial obscure science, argue that any natural variation proves their point and cannot integrate data from all disciplines of science to get an overview of the holistic workings of planet Earth. Some even change the primary data and use "corrected" data.

Creationists are quite happy to read only the creationist literature, controlled by their masters. Climate "scientists" are quite happy to only read the climate literature that is tightly controlled by the catastrophic climate club set up to maintain a consensus and keep the funds rolling in. Journals, editorial boards and peer reviewing are controlled by climate "scientists". Both groups do not display a broad knowledge, have no well-read polymaths and can only stick to their group-think message. This is especially the case with media networks that stick to the safety and ignorance of group-think. This group-think is not helped when practitioners have no background in science.

At the rare times when there is public debate, both groups make expressions of faith and morality, paint doomsday scenarios and they cannot present the simple science that underpins their dogma. The arguments of both groups fail when confronted with geology, evolution and extinction of life and sea level changes, all of which are dealt with in this book. Time unstitches both the climate industry and creationists. Creationists use fraud to try to show that radioactive dating of rocks demonstrates that planet Earth is only a few thousands of years old rather than 4,500 million years old. Climate catastrophists also totally ignore time. Geology is excluded from climate debates because it shows that past changes in climate have been greater and quicker than anything measured today and are not related to atmospheric CO_2. Climate catastrophists only focus on the last 20 years to predict what will happen hundreds of years into the future.

Climate catastrophists, like creationists, normally don't collect the primary data themselves. They massage the data of others and, in the case of climate catastrophists, use intricate computer models to try to frighten their flock. Inconvenient measurements are conveniently ignored. For example, a large number of green left environmental activists ignore the data that shows there has been no increase in global warming for more than 18 years. Courts often exclude data as being inadmissible leading to legal decisions that are incomprehensible by the person whom the courts allegedly serve. Science does not. Creationists ignore geological data whereas various climate centres (e.g. Hadley) ignore large bodies of data (e.g. Russian temperature data, palaeoclimates, geological history of CO_2) that don't back their preconceived story. Others (e.g. CRU) keep their data in such a chaotic mess that nothing can be checked. Messages from God abound. Any variation in a natural process over the last 100 years is, of course, due to the action of humans or God, depending upon whether the speaker is a climate catastrophist or a creationist.

Even when shown to be scientifically wrong, both groups still cling to their treasured theories. Creationists and the climate industry have lectures, public meetings and conferences that are closed shops. No one who could challenge their views in front of others is invited or can attend. This was the case with the preparation of the papal Encyclical.

Both groups need something to fear and become catatonic if someone tries to take away their fears, no matter how irrational. Both ignore history. With the climate industry, history shows that previous warmings have been longer and more intense yet non-catastrophic than the very slight late 20th century warming. The past warmings could not possibly have been driven by industry belching out CO_2 because, unless all of history is wrong, there was no heavy industry in past periods of warming such as the Minoan Warming, Roman Warming and Medieval Warming.

History is also pretty cruel to creationists because it appears that there were whole empires, towns and villages on planet Earth well before their Earth was created, that many communities were not wiped out by a great global flood and that all sedimentary rocks and fossils could not possibly have formed in this flood. Both climate catastrophists and creationists just do not accept history and even try to change history. Creationists try to change dates such to make history agree with their view of the world; similarly the Medieval Warming was simply removed from the record by Michael Mann, the IPCC and green left environmental activists.

For both groups, carbon 14 (C^{14}) causes huge problems. Creationists are a little more sophisticated than the climate industry and use arguments about the accuracy of C^{14} dating to validate their view of history. They have trawled through the scientific literature and found the occasional published C^{14} radioactive date that is wrong and they have attacked the methodology of radioactive age dating. The climate industry simply ignores C^{14} evidence that shows changes in solar activity and cosmic rays resulting in increased cloud cover and cooling.

Both climate catastrophist and creationist dogma can be easily destroyed with a traverse through a thick section of layered sedimentary rocks which show that planet Earth is old, that sea level constantly changes, that climate rapidly changes from icehouse to greenhouse conditions, that climate is never static, that planet Earth for most of time has been warmer than now and that life survives these great changes. This is not new. It was done in the 18th century and has been validated thousands of times. But, why should either group use empirical evidence when belief is a matter of faith? This, of course, is a waste of time. No matter how many times the climate catastrophist and creationist dogma is shown to be wrong, their dogma, faith, financial interests and career prospects just do not allow them to abandon a treasured and lucrative theory.

I somewhat modestly claim that I am an expert on Noah's Ark. It is a dubious honour but someone has to carry this immense burden. This expertise is supported by various writings, broadcasts, television programs at Mt Ararat (ABC) and the high Atlas Mountains (BBC) and litigation regarding Noah's Ark. The Noah's Ark message is that God destroyed most life on Earth with a massive worldwide flood because we were not behaving ourselves. The good guys and a few species carried on the Ark survived because they were sensible enough to fear and listen to God. Sinners and most life died.

We see the same moralistic thinking among climate catastrophists. We humans are evil, there are messages sent in the atmosphere, life and oceans by Gaia and we need to change our ways otherwise we are finished. Again, fear and guilt are the weapons used to try to force us to change our ways. If we don't pay up and change our life, then we will fry-and-die, sea level will rise and there will be massive extinctions of life. There might just be space for us on planet Ark but a berth will cost. This Noachian religious remnant underpins much of the green left environmental atheistic movement and has been embraced by scientists profiting from the catastrophist climate cause. The Ark is the environmentalists' religious symbol for bio-diversity. Both groups redefine science, have a very selective use of data are unable to revise their cherished theories and become aggressive when their opinions are challenged with contrary evidence. There is a chronic aura of insecurity in both groups who feel that the Earth was once perfect and static in either the Garden of Eden or pre-industrial times. Both groups believe humans have destroyed the world by sin.

This shows that both groups are essentially anti-human. Both movements have all the trappings of humourless Western funda-mentalist cults which intertwine their theology with a world-view replete with morality and both argue that science underpins their worldview. Creationists have little interest in theology, just as the cli-mate industry has little interest in the environment. It is a game of power. The power games involve mutual support groups, shutting out of dissidents and threats to apostates of total destruction, either as eternity in Hell or the destruction of a career. The climate in-dustry, like creationism, ignores data, creates data *ex nihilo*, "adjusts" data, selectively uses data and cooks the books, albeit in different ways. This unifying thread of deceit ties both groups together. Why is there the necessity for both groups to "adjust" raw data?

When I was challenging the creationists in public, I was the darling of the left. They incorrectly believed that I was attacking Christianity. I was not. I was attacking fundamentalists who exploited Christianity to claim that science underpins their scientifically incorrect world-view. I attacked the creationists because their scientific method was erroneous, deceptive and fraudulent. All mainstream Christian religions supported me including many priests and laity of the Catholic Church. The Sydney Catholic Education Office saw the dangers of creationism and was a great supporter of my anti-science activities.[552]

I attack the climate industry because their scientific method is erroneous, deceptive and fraudulent. I also attack the climate industry because they exploit people's yearning for spirituality by claiming that science underpins their scientifically incorrect worldview. Now the left attacks me because their non-scientific view of the world is anti-industrial and anti-human. It is interesting that ABC Science Show presenter Robyn Williams was a great supporter of my attacks on creationism (and, together with Archbishop Peter Hollingworth, wrote a Foreword to *Telling lies for God*). He agreed with my view that the scientific methodology of creationism was erroneous, deceptive and fraudulent. Mr Williams now attacks me because I argue that the scientific methodology of the green left environmental activists is erroneous, deceptive and fraudulent. I attack both creationism and the climate industry because they abuse science.

The climate catastrophist movement has been embraced by much of society because they are either not scientifically educated or they need to believe in something. That does not mean they are not smart. The climate catastrophist movement is the new fundamentalist religion of the West that has replaced Christianity and has many elements of the replaced religion (sin, indulgences, salvation, worship of a higher authority). However, the new environmentalist religion is atheistic, anti-human, vacuous and secular with no music, history, literature, compassion, charity, schools, coherent philosophy or deep thinking. What is surprising is that the Pope was not advised well enough about the new green religion that increases poverty and destroys the environment.

552: Price, Barry 1987: *The bumbling, stumbling theory of creation science*. Catholic Education Office

Just scratch the surface of a creationist or a climate catastrophist, and you will find a very angry person. Just scratch the surface of a Green Party politician and you'll find the same. If you don't believe me, look at the green left environmental activist Tweets, blogs, Facebook postings and broadcasts. There is an army of vulgar ill-educated angry people out there who cannot mount a logical argument in 140 characters and just use abuse against anyone who may differ. This is the mechanism of closing down the argument. One does not need to think, analyse, criticise or have knowledge. Argument is allegedly won by the loudest and most vulgar abuse. They are not debaters, they are haters. The green left environmental activists promote intolerance, totalitarianism and loss of freedoms. Are these the people with whom the Pope identifies?

I see very little difference between creationism and climate catastrophism. We could satisfy our spiritual needs with music, history, literature, thinking and a coherent spiritual philosophy rather than a vacuous totalitarian intolerant environmental quasi-religious fad to save the world.

In our free democratic society, I am intolerant of intolerance.

4

THE MORAL CASE

THE GOOD OLD DAYS

Winning the lottery

The Pope longs for past times,[553] when times were apparently better. They weren't. I'm old enough to remember living in what some young people think were the good old days. I have no memory of picking flowers, dancing in the fields, singing arm-in-arm and being happily at one with nature. I did love picking fruit off trees and eating it on the spot. This was because it was stolen from neighbours' fruit trees and the evidence had to be destroyed as quickly as possible. However, I remember frugal times. I remember grandparents, relatives and family friends who were ancient to me but were only in their 60s and 70s and had medical ailments that could easily and cheaply be treated today. I am now older than when many of these folk died.

Most of these people never finished school and I was the first generation in my family to finish school, read for a university degree, obtain a higher degree and work using my brain rather than brawn. I never knew economic depression, a world war or hunger yet I have known refrigerators, washing machines, gas/electric stoves and ovens, air conditioning, television, supermarkets, frequent national and international travel, a diversity of good quality fresh food and outstanding local and instantaneous global communications. I now don't have to book an international telephone call a few days in advance.

553: *Laudato Si'*, Paragraphs 46, 47 and 175

I just do it. My grandparents and their parents didn't know of such things.

The Pope is concerned about *"fertilizers, insecticides, fungicides, herbicides and agrotoxins in general"*[554] yet it is these chemicals plus atmospheric plant food that have enabled more than seven billion people to be fed and, as a result of far more efficient agriculture, the Earth is revegetating at a greater rate than it is devegetating.

Only a short time ago in the Western world, highly contaminated waste and sewage water was also drinking water. People died like flies, especially women and children, and God help you if you had an infection, scratch or wound. Average longevity was less than 30 years. It was reticulated potable water and separate waste water systems that saved more lives than any Pope, policy or politician. The UN's Millennium Development Goals 2014 Report[555] states that *"access to improved drinking water source became a reality for 2.3 billion people"* over the past 20 years and that *"the target for having the proportion of people without access to an improved drinking water source was achieved in 2010, five years ahead of schedule"*. People still die from poisons, pathogens and poo in contaminated water but the quality of available water[556] is not *"constantly diminishing"*. UN and World Bank studies show the exact opposite. It might have been more prudent for the Pope to state that there have been great advances but there is still much to achieve.

Don't give me the good old days. They weren't. I won the lottery of life: I was born in post-World War II Australia. Many New Zealanders, Canadians and Americans of my age would feel the same. I was born into a rapidly growing economy wherein, over five decades, the population became wealthy, secure and well fed because of hard work, freedom, free trade, education, creativity, flexibility, property

---

554: *Laudato Si'*, Paragraph 20
555: http://www.un.org/millenniumgoals/2014%20MDG%20report/MDG%20 2014%20English%20web.pdf
556: *Laudato Si'*, Paragraph 30

rights, lack of oppressive traditions, lack of stifling regulation and careful husbandry of natural resources. We did not become modern wealthy Australians with excessive debt, foreign aid, laziness, entitlements, drugs, over-regulation and excessive welfare. We did it from scratch ourselves and with the help of immigrants. No one else did it for us. The Pope would better serve the global community if he used the examples of the growth of New World democratic free market countries to show the developing world the path out of poverty. This would be a better service than espousing policies of dependency and principles of socialism.

The proof of the pudding is in the wall of humanity trying to illegally enter New World countries such as along the southern border of USA. Why do Hispanics want to illegally enter the USA? Because it is a far better place than their homelands. Maybe the Pope could better serve these poor people by promoting the benefits of free markets, capitalism, democracy and freedom rather than promoting socialism. Individuals and countries do not instantly become wealthy (unless one is a Russian oligarch, African despot or union leader) and, over decades to centuries, the New World economies used market economics, tapped into human creativity and constantly delivered innovative solutions in a constantly changing world. This is exactly what is happening in Asia and the Indian sub-continent today.

Doomsday predictions

The Encyclical is a depressing document with doomsday fear and posing of seemingly insoluble problems. However, this is done in the absence of history. By every measure, life is better now than it has ever been. Life expectancy is higher, infant mortality is lower, daily calorific intake is higher and more people have access to education, clean water, electricity and housing. Probably the best measurement of the current good times is life expectancy. Compared to the Third

World, in the Western world we are now well off and suffer the curse of affluence.

In 1800 AD, although there were only one billion humans on Earth, global life expectancy was about 24 years. The world's population had doubled by 1927 and people could expect to live twice as long. At present, there are more than seven billion people on Earth and average longevity is 69 years. Life expectancy in the poorest nations now is far better than in the richest nations 150 years ago. The average life expectancy of a child born today in the Western world is more than twice as long as in my grandparents' day a little more than century ago.

In the Third World, it is the same story. For example, in 1906 average life expectancy of a Bangladeshi was 21.5 years. The average Bangladeshi child born today can expect to live to 75 years old. This has been achieved by doing the exact opposite of what the Pope has advocated yet he promotes the irrational fear that we are on the eve of destruction.

The Pope writes:

> *Doomsday predictions can no longer be met with irony or disdain.*[557]

and

> *There are regions now at high risk and, aside from all doomsday predictions, the present world system is unsustainable from a number of points of view …*[558]

For thousands of years there have been global doomsday predictions. Not one has been correct. If just one of the zillions of end-of-the-world scenarios had occurred, then we wouldn't be here today. Complex mathematical forecasting systems have been shown to be far better than expensive computer climate models used by the

557: *Laudato Si'*, Paragraph 161
558: *Laudato Si'*, Paragraph 61

IPCC but such a sober look at the world does not see the light of day in the popular sensationalist media.[559] Scientists, religious leaders, your near-neighbour nutter and even domestic animals all have made predictions about disasters.[560] Environmentalists have now taken over the field that was once the preserve of colourful characters.[561] Global warming is just one of thousands of environmental dooms-day scenarios, justified by dubious scientific claims, mathematics and statistics.

Modern environmentalism

We are all environmentalists. No one wants to see the atmosphere, soils and waters polluted. We can all do something about pollution and the wealthy Western world has cleaned itself up over the last 50 years. This is because of wealth and culture. A few decades ago, some of the Mediterranean countries were cesspits of garbage (e.g. Spain, Italy, Greece). They are more environmentally aware today and are still changing the throw away littering culture. Most develop-ing countries have a culture of throwaway littering and will need to be wealthier and have a change in culture before they are environ-mentally comparable with the West.

Fairfield Osborn can be accredited with starting the modern en-vironmental scare movement and reviving Malthusianism.[562] True to form, he was a eugenics supporter and prominent Aryan enthusiast. The 20th century environmental movement grew out of eugenics,

---

559: Green, K. C. and Armstrong, J. S. 2007: Global warming forecasts by scien-tists versus scientific forecasts. *Energy and Environment* 18: 997-1021

560: Randi, James 1995: *An encyclopedia of claims, frauds and hoaxes of the occult and supernatural.* St Martin's Press

561: On the basis of biblical interpretations, Harold Camping publicly predicted the end of the world 12 times, the last date was 21 October 2011. His end of the world came on 15 December 2013. His critics had been right in stressing Matthew 24:36 ("of that day and hour knoweth no man").

562: Osborn, Fairfield 1948: *Our plundered planet.* Faber and Faber

Malthusian doomsday scenarios, Nazi movements and totalitarianism.[563] It still retains some of these elements. Paul Ehrlich[564] gained
great recognition by attempting to scare us witless by suggesting that
"the battle to feed humanity is over" and that "hundreds of millions
of people would starve." Ehrlich was a follower of eugenics enthusiast William Vogt[565] and his solution to perceived population problems was to add chemicals to staple foods to create sterility. Books by
Paul Ehrlich derive from Malthusian ideas, the eugenics movement
and communism (all of which underpin the modern green left environmental movement) and espouse the centralised control of every
aspect of life.

Ehrlich stated that the world would run out of resources. By contrast, Julian Simon[566] suggested that increasing wealth and technology
make more resources available and that increased population is the
solution to resource scarcities and environmental problems because
people are creative and free markets allow creativity. History shows
that this is exactly what happened. Simon was right, the doomsdayer
Ehrlich was wrong. Simon and Ehrlich had a bet in 1980. Ehrlich
bet that the commodities copper, chromium, nickel, tin and tungsten
would increase in price from 1980 to 1990 during a time when the
world population would increase by 800 million and when Ehrlich
thought that raw materials would become scarcer. All five commodities decreased in price, Ehrlich lost the bet and paid up. Ehrlich was
wrong.

When doomsday scares are investigated they are shown to be totally wrong, result in poor public policy and cost a huge amount of
money that could be spent dealing with real rather than fraudulent,

563: Ray, Dixy Lee 1993: *Environmental overkill: Whatever happened to common sense?*
Regnery Gateway
564: Ehrlich, Paul 1968: *The population bomb.* Sierra Club/Ballantine Books
565: Vogt, William 1948: *The road to survival.* William Sloan Associates
566: Simon, Julian 1981: *The ultimate resource.* Princeton University Press

hypothetical or exaggerated problems. We are getting used to Ehrlich's failed predictions. Followers are comforted by the fact that they have a messiah and hence do not need be bothered with the trivia such as reading history, understanding science and thinking critically and independently. Have I got news for Paul Ehrlich and his followers. As many have shown,[567,568] the world is far cleaner, healthier and wealthier than in 1968. Ehrlich was wrong. Again. Like so many green left activists, he cannot face the prospect of being demonstrably wrong. I don't know what cave Ehrlich has been hiding in for decades but in 2009 he suggested that the main flaw in his book was that it was too optimistic!

The Club of Rome, a global doomsday green left think-tank, commissioned a book[569] which predicted that the world would run out of various non-renewable resources in the 1980s and 1990s and that environmental economic and societal collapse would follow. The computer predictions were based on exponential economic and population growth using only five variables. The world didn't run out of resources but may have run out of common sense. Nothing has changed with computer predictions. The Club of Rome was wrong. Some 30 years later, the authors produced an updated version.[570] Surprise, surprise, the updated doomsday predictions have also been shown to be wrong.

Some doomsday predictions are self-fulfilling for the wrong reasons. Rachel Carson[571] was primarily responsible for the 1972 ban-

567: Lomberg, Bjorn 2001: *The skeptical environmentalist.* Cambridge University Press

568: Booker, Christopher and North, Richard, 2007: *Scared to death. From BSE to global warming. Why scares are costing us the earth.* Continuum

569: Meadows, D. H., Meadows, D.L. and Randers, J. 1972: *Limits to growth.* Universe Books, 1972

570: Meadows, D. H., Randers, J. and Meadows, D. 2004: *Limits to growth: The 30-year update.* Chelsea Green Publishing

571: Carson, Rachel 1962: *Silent spring.* Houghton Miffin

ning of DDT as a control on malaria on the grounds that it killed birds and caused a cancer epidemic. Both of these unsupported claims were later shown to be untrue. During World War II in the Pacific and SE Asia, DDT reduced loss of Allied troops to malaria leading to more troops to do the fighting. The World Health Organisation reversed the ban on DDT in 2006. Between 1972 and 2006, at least 50 million people unnecessarily died of malaria. Most of the deaths were in Third World countries and most of the deaths were children. This is the track record of environmentalism. Carson has blood on her hands and the order of magnitude of killing puts her on the same pedestal as Mao Tse Tung, Stalin and Pol Pot.

Carson also claimed that acid rain was devastating German forests and this was repeated *ad nauseam* by the German Greens in the 1980s. The German forests have been so devastated that they have expanded! Al Gore suggested that the ozone hole was making rabbits and salmon blind. I'm not so sure why rabbits would want to look at salmon or salmon need to look at rabbits but both animals didn't go blind.

In 2013, the EU imposed a ban on using neonicotinoid insecticides because the honey bee population was declining. The decision was based on faulty science, the honey bee population was not declining, there were an additional 900,000 hives in Europe and wild bees who are most likely to come in contact with neonicotinoids are thriving. As a result of the ban, oilseed rape (canola) production decreased by 7 to 20%. This was not the precautionary principle in operation. It was hysteria based on flawed science.[572]

Green left environmental activists have successfully prevented the farming of golden rice, genetically modified rice with vitamin-A precursor traits introduced from maize. This rice was specifically developed to tackle the problem of vitamin-A deficiency in many poor

572: *The Wall Street Journal, Europe*, 23 July 2015, http://93.114.44.238/view-topic.php?f=19&t=1192145

countries. Hundreds of thousands of children die each year from vitamin-A deficiency. GM golden rice has been ready to save lives for years but every step is opposed by Greenpeace. In Bangladesh, farmers risk their own health and spray egg plants with insecticide up to 140 times a season because the insect-resistant GM variety of the plant is fiercely opposed by environmentalists. After 20 years and billions of meals, there is no evidence that GM foods damage human health. Only the opposite has been seen in this 20-year period: GM foods save lives and have environmental benefits.

The environmental movement has repeatedly denied people access to safer and cheaper technologies and forced them to rely on dirtier, riskier or more harmful technologies. The movement exploits human's fears and has exaggerated the dangers of climate change. These claims have been shown to be wrong.

The word belief is used in religion and politics. It is not a word of science. Over the last 30 years it has been used by environmentalists who claim a scientific basis for their beliefs. It has been argued that people form their beliefs from emotion and sentiment and that rational justification of beliefs are constructed after the belief has been accepted.[573] The scientific evidence for human-induced global warming by CO_2 emissions is incredibly weak and the basis of a belief in global warming is emotional. The emotional core of the global warming belief is a fear of modern technology. In many places in the Encyclical, the Pope shows he is wary or frightened of modern technology and/or change.[574]

When objections are raised about doomsday projections and the green left global warming mantra, there is no attempt to meet the objections with scientific arguments. Critics are attacked and mar-

---

573: Pareto, Vilfredo 1968: *The rise and fall of elites: An application of theoretical sociology*. Transaction Publishers (Translation of 1901 original)
574: *Laudato Si'*, Paragraphs 9, 16, 20, 54, 60, 102, 106, 107, 108, 109, 110, 112, 114, 131, 132, 136, 165 and 172

ginalised, again showing that a belief in global warming is emotional. Honest debate on environmental concerns is very rare. Attempts at debate are met with smear, hysteria, claims of settled science etc and every attempt is made to shut down debate. The science behind the global warming scare is so weak that by any measure, it should be rejected. The end result of this environmental scare translated as public policy results in unemployment, premature death, poverty, unemployment, a loss of freedoms, massive futile government expenditure and mandated high cost energy sources which distort and disrupt economics, increase debt and damage economic growth. There is a bottom line.

A better human environment

Due to better medicine, health care and nutrition, infant mortalities have fallen. In the late 19th century, mortality globally of children under five was 40%. It is now 6% and falling. Mortality rates in Western countries are extremely low. The Pope's Encyclical is suggesting that we go back to these days when curable fatal diseases were rampant. The decreased mortality rates have played a major role in population increase and longevity. So too has food. Crop yields per hectare and food consumption *per capita* have been increasing for a century and have never been higher, despite the population increase. Enough food is now produced for every person on Earth to consume 3,500 calories per day and there is no need now for anyone to starve. Furthermore, this food is produced from less land than decades ago due to fertilisers, herbicides, insecticides and genetically-modified crops. This has led to an increase in forest lands on planet Earth.

However, the Pope wants us to go back to simple agrarian times with small community plots of land,[575] to use no fertilisers[576] and to

---

575: *Laudato Si'*, Paragraph 67
576: *Laudato Si'*, Paragraph 20

avoid genetically modified crops and cotton.[577] This can only lead to mass social disruption, starvation, depopulation and the pushing of billions of people into poverty. We have been there before and it is not a pretty sight. Starvation still occurs because of wars, drought, political upheavals, incompetence, tribalism, unreliable transport systems and despotism.

Assessments of the impacts of global warming give short shrift to the benefits of global warming, whether driven by human activity or whether natural. Mortality data from many countries,[578,579] regions[580] and cities[581] with cold, temperate,[582] subtropical,[583] tropical[584,585] and arid[586] climates shows that mortality is substantially higher in the cold months than in the warm months. Jack Frost and the Grim Reaper are good mates. The only good thing about cold weather is that it

---

577: *Laudato Si'*, Paragraphs 131, 132, 133 and 134

578: Guo , Y. *et al.* 2014: Global variation in the effects of ambient temperature on mortality: A systematic evaluation. *Epidemiology* 25: 781-789

579: Vardoulakis, S. *et al.* 2014: Comparative assessment of the effects of climate change on heat- and cold-related mortality in the United Kingdom and Australia. *Environ. Health. Perspect.* doi: 10.1289/eph.1307524

580: Falagas, M. E. *et al.* 2009: Seasonality of mortality: the September phenomenon in Mediterranean countries. *Canad. Med. Assoc. Jour.* 181: 484-486

581: Yi, W. and Chan, A. P. 2014: Effects of temperature on mortality in Hong Kong: a time series analysis. *Internat. Jour. Biomet.* 58: 1-10

582: Berko, J. *et al.* 2014: Deaths attributed to heat, cold, and other weather events in the United States, 2006-2010. *National Health Statistics Reports* 76: 1-16

583: Wu, W. *et al.* 2013: Temperature-mortality relationship in four subtropical Chinese cities: A time series study using a distributed lag non-linear mode. *Sci. Tot. Envir.* 449: 355-362

584: Burkart, K. *et al.* 2011: Seasonal variations of all-cause and cause-specific mortality by age, gender, and socioeconomic condition in urban and rural areas of Bangladesh. *Int. Jour. Equity Health* 10: 32

585: Egondi, T. *et al.* 2012: Time-series analysis of weather and mortality patterns in Nairobi's informal settlements. *Glob. Health Action* 5. Doi 10.3402/gha.v5i0

586: Douglas, A. S. *et al.* 1991: Seasonality of disease in Kuwait. *Lancet* 337: 1393-1397

forces politicians to keep their hands in their own pockets. In many parts of the world, retirees move to warmer climates (e.g. Sun Belt of USA; south of France or Spain by British retirees). The evidence is there. Warmer weather does not increase mortality rates. Unless, of course, you are silly enough to believe the models of climate "scientists."

Not only have longevity, infant survival and health increased, the pocket has become heavier. The rate of *per capita* income has outstripped population growth. Over the past two centuries population has increased by a factor of seven whereas *per capita* income has increased by a factor of 90 times from $100 to $9,000. Although there are currently a few short-term exceptions (e.g. Russia, Greece), economies are expanding, productivity is increasing, poverty is decreasing, pollution is declining and political freedoms are becoming more widespread. The human race has never been so well off and we are blessed to live on Earth today rather than hundreds or thousands of years ago.

We are told that if human emissions of CO_2 are not curbed soon, then there is a threat to humanity. The green left environmental activist group 350.org wants CO_2 reduced from the current 400 ppm to the 1988 level of 350 ppm. Since 1988, global GDP has increased 60%, infant mortality has decreased 48%, life expectancy has increased 5.5 years and the poverty headcount has dropped from 43% to 17% despite a population increase of 40%.[587] The past was not very pretty and the world has never been better as shown by crop yields, food production, hunger, access to clean water, biological productivity, life expectancy and standard of living.[588] It seems that the climate catastrophists ignore the fact that human ingenuity can solve most problems and has made the planet a better place.

These are great advances, bought to the planet mainly by coal.

587: http://databank.worldbank.org/data/databases.aspx
588: Matt Ridley, 2010: *The rational optimist. How prosperity evolves.* Harper Collins

The Pope wants to bring us back to these times by putting limits to growth, *"retracing our steps before it is too late"* and *"put aside the modern myth of material progress"* by slowing down mechanisation, turning down the heating and putting on warmer clothes.[589] The four billion poor people on planet Earth have a different view. They want the cheapest form of reticulated electricity for cooking, heating, air conditioning and employment and this comes from the burning of coal. This is what we did in the Western World to rise from poverty to our current wealthy state and the Pope is advocating that we keep most of the world in grinding poverty and that we in the West should join them.

In the early 19th century, the Industrial Revolution delivered people from the grinding poverty of the peasant class. The middle classes appeared and all of this was due to science that underpinned technology. The use of this technology was profit-driven, a process that seems to unnerve the Pope.[590] The march of technology, progress and economic growth should be lauded by the Pope as it brought humans from poverty, gave hope, education and wealth and the ability to improve our lives on Earth.

This increase in both the length and quality of life is due to technology, insecticides, herbicides, reticulated water, free trade, transport via fossil fuels, refrigeration, air conditioning heating, reticulated electricity and medicine. Notwithstanding, we need to put air pollution into context. About half the deaths from air pollution are from smoky indoor fires in poor countries. Cheap reticulated coal-fired electricity would solve this problem, as it did in Western countries.

Southern Africa is gearing up to build a dozen or so coal-fired power stations contrary to Western green left environmental pressure. Why did sub-Saharan Africans ignore Western pressure? Be-

---

589: *Laudato Si'*, Paragraph 211
590: *Laudato Si'*, Paragraphs 9, 16, 20, 54, 60, 102, 106, 107, 108, 109, 110, 112, 114, 131, 132, 136, 165 and 172

cause they want a better life. South Africa, Botswana, Malawi, Mozambique, Namibia, Zambia, Zimbabwe and Tanzania all have an abundance of low-sulphur low-ash high quality coal. Proven coal reserves in South Africa are about 32 billion tonnes of economically recoverable coal and the resources are probably an order of magnitude higher. Two massive 4,800 MW plants are under construction and there are proposals for building many smaller plants in the order of 300 MW.[591] Southern Africa needs cheap electricity, needs to profit from its vast coal resources and will prosper from building coal-fired power stations.[592] This is the first step out of widespread poverty.

Landlocked Botswana needs to diversify from its economic dependence on diamonds and tourism and create wealth from its vast coal deposits. Developing coal resources will require capital and infrastructure. Tanzania generates 1,000 MW, requires 2,000 MW and the GDP is rising by 7% *per annum* hence a great increase in the amount of electricity is needed. As with every other place in the world, demand for electricity rises with income. Tanzania aspires to have an electrical energy mix of 30% gas, 30% hydro and 30% coal.

Without cheap electricity, prosperity can only be a dream. Nearly 730 million Africans rely on wood, twigs, leaves and dung for cooking and heating and 620 million have no access to electric lights. That's two-thirds of the population. In sub-Saharan Africa, 40% of people in urban areas have no electricity and 85% of people in rural areas have no electricity. Globally, 1.3 billion people don't have access to electricity and 2.7 billion people rely on wood, twigs, leaves and dung for cooking and heating which causes harmful indoor air pollution. Without cheap electricity, production cannot be increased, goods can't get to market, vaccines cannot be kept in refrigeration

591: South African Development Community, www.sadc.int
592: *The National, Business*, 25 April 2015

and hundreds of millions of young people cannot study after the Sun goes down.

Without education, health and cheap employment-producing energy, Africans cannot escape from poverty. The papal Encyclical deals with education, health and poverty in many places yet does not consider that cheap energy is the solution to such perennial problems. Does the Pope want to keep Africans in such poverty? What do the Pope's African bishops have to say on the papal Encyclical that condemns almost a billion of their own people to eternal poverty? The International Energy Agency[593] stated that "increasing access to modern forms of energy is crucial to unlocking faster economic and social development in sub-Saharan Africa." The greens want to prevent Africans from using the cheapest form of energy: Coal. The greens are knowingly keeping people in poverty and killing people.

Most of us would share the Pope's desire to reduce poverty. The environmental and especially the energy policies in the Encyclical will achieve the opposite effect. There are numerous examples from the last 200 years to support this conclusion.

In an amusing Orwellian capture of the language, those that call themselves progressives want to push us back to the times when we died like flies from starvation and disease.

The human fingerprint

Cores from ice sheets show when the Romans smelted lead in Iberia, when there were major land-clearing events and forest fires, when trace elements were added to the atmosphere during the Industrial Revolutions and when radiogenic isotopes were released in modern times.

The future geological record will show a unique sediment layer defining when modern humans existed. The layer will show that

593: Africa Energy Outlook, International Energy Agency 2014

shallow water lake and continental shelf sedimentation had increased and this layer contained an increased content of wind-blown dust. This layer will be chemically defined by complex synthetic organic molecules such as plastics, polycyclic aromatics, polychlorinated biphenyl, insecticides, fungicides and herbicides. It will have a trace element content slightly enriched in sulphur and base metals from smelting operations, a high phosphorus, nitrogen, mercury, arsenic, intermetallic, platinoid and rare earth element content and the presence of synthetic isotopes from nuclear bombs. The layer will be thin and will be best preserved in deep water oceanic sediments.

These global pollution event will essentially record when there was growing industry in the developing world and when countries were at their maximum development, the spectrum of the chemical fingerprint increases. When we look at the times and scale of geological processes, this human fingerprint will be a minor curiosity with larger fingerprints from volcanoes, asteroid impacts, climate change and crustal evolutionary changes dominating the strata.

Get used to it. We humans are insignificant.

Numbers don't lie

What is energy

Time to bore the pants off you. The First Law of Thermodynamics states that the total energy of an isolated system is constant and energy can be transformed from one form to another but cannot be created or destroyed. The First Law is about the quantity of energy. The Second Law of Thermodynamics is about the quality of energy and essentially states that as energy is transformed, more and more is wasted and there is a natural tendency of any isolated system to degenerate into a much more disordered state. Getting old is a good example of the Second Law of Thermodynamics. We have known the essence of these laws for almost 200 years.

The Earth has three principal sources of energy. What we do as humans is convert one form of energy into another. The first source of energy is from very deep within the Earth and derives from the core and radioactivity. The core of some extraterrestrial bodies has changed from liquid to solid and frozen (e.g. Mars, Moon) whereas the Earth's core is still molten and retains its heat. This heat is transferred to the mantle and crust mainly by convection through almost solid rock. Plumes of convected material move continents and plumes partially melt upper mantle rocks. These melts are buoyant and heat the Earth's crust.

Heat flow maps show that most of the deep heat from the mantle is in the circum-Pacific 40,000 km-long Ring of Fire, the Mediterranean-transAsiatic Belt and the mid ocean ridges. Some of this heat is used in geothermal power stations such as in New Zealand, Philippines, Italy and Iceland. About 0.1% of the world's electricity is generated from geothermal systems.

The decay of potassium 40 (K^{40}), uranium 238 (U^{238}), uranium 235 (U^{235}) and thorium 232 (Th^{232}) produces heat over an extraordinarily long time. It is this heat that keeps the Earth's mantle hot over geological time. If radioactive minerals are concentrated by geological processes near the surface of the Earth, they can be mined.

Natural uranium ores contain 99.284% U^{238} and 0.711% U^{235}, the ore is processed to make yellow cake[594] which is sold at the mine gate and is processed elsewhere into a product that is enriched to about 20% U^{235} for use in a reactor. Accelerating the decay of U^{235} in a reactor produces breakdown products comprising daughter isotopes of lighter elements, particles and heat. The reactor heat is used to drive steam turbines that spin magnets to generate electric-

---

594: Uranyl hydroxide, triuranium octoxide, uranium dioxide, uranium trioxide, uranyl sulphate, sodium-para uranate, uranium peroxide, ammonium diuranate or sodium diuranite are loosely called "yellow cake". Different mine processing plants produce different yellow cakes.

ity. The daughter isotopes can be reprocessed for further heat generation and shields around the reactor stop particles leaving the reactor. About 5.1% of the planet's total electricity is generated from nuclear reactors.

The second principal source of energy is stored as carbon-rich materials in the top few thousand metres of the Earth. This energy is in the form of stored solar energy that is now combustible carbon-rich materials such as coal, oil, gas, tar sands and oil shales. This energy derives from sunlight, is stored in life and compressed into gas, oil or rock. The chemical process of oxidation converts carbon compounds into CO_2 and H_2O. This gives out heat that can be used to spin a turbine with steam or drive an engine crankshaft. This process of hydrocarbon combustion drives the world with electricity, transport and chemicals used in everyday life.

The third principal source of energy on Earth is extraterrestrial. Solar radiation can be used directly to generate electricity. Radiation from the Sun is captured by the atmosphere and oceans as heat. Water evaporates from seas and the oceans and falls as rain. Lunar tides and the gravitational flow of rainwater from highlands can both be used to generate electricity. The transfer of atmospheric heat results in wind and storms. If wind flow is stopped, the energy can be used to generate electricity but reduction of wind velocity can create rain shadows behind turbines. The Sun and rain stimulate plant growth and biomass can be burned to produce heat to make steam to generate electricity. Without the Sun, there is no life on Earth (with the exception of a few weird bacteria). The third source of energy is most commonly referred to as green energy. It is not so green. It reduces the supply of food, water and available energy to life on Earth, kills wildlife and consumes large amounts of fossil fuels for manufacture, construction, maintenance and backup.

If the greens were concerned about the planet they would support the use of fossil fuels because the amount of land used per unit energy is small (compared to solar, wind, hydro or biomass), the combustion processes for fossil fuels and biomass are the same, combustion adds the essential ingredients of life to the atmosphere (CO_2 and H_2O) and fossil fuel burning does not kill wildlife. If the greens were really concerned about the land area used, emissions and damage to wildlife, they would also be great advocates of the nuclear industry.

U^{235} has the highest energy density of all forms of energy used to create electricity. If all the energy you needed to live your three score and ten years derived from U^{235}, the end result would be half a cup of spent fuel. To make 1,000 kWh of electricity, you need 0.03 g of pure U^{235}, 265 litres of oil, 379 kg black coal, 1,000 one metre square solar panels or one 600-kilowatt wind turbine operating flat out for 1.5 hours.

Energy use

In 2013, the BP Statistical Review of World Energy showed that about 87% of global energy consumed came from fossil fuels for transport (oil), heating (gas) and electricity and smelting (coal).[595] This figure was the same in 2003. In Western countries, there has been a market-driven decarbonisation with a higher proportion of gas burned. Gas burning releases less CO_2 than burning coal. The slight decarbonisation was not driven by wind or solar power. In 2003, nuclear contributed to 6% of all energy consumed. Now it is 5% and the US Energy Information Administration[596] forecast it will rise to 6.7% in 2025.

595: http://www.bp.com/content/dam/bp/pdf/Energy-economics/statistical-review-2014/BP-statistical-review-of-world-energy-2014-primary-energy-section.pdf
596: http://www.ela.gov/

Energy history

Up until the 17th century, feet and wood were used to produce energy. A little energy came from water. People and animals died like flies and land was completely deforested for the making of iron and glass. In the 18th century, wood was the main source of energy and forests were destroyed at a very rapid rate. In fact, the area of forests now is far greater than it was in the 18th century. In the 19th century, coal was the main form of energy. This was a period of great innovation. It was the high energy density of coal that drove this innovation and the Industrial Revolution. In the 20th century, it was oil that drove industrialisation, transport and trade and coal was used for smelting and coal-fired power stations.

Once there was abundant cheap coal-fired electricity in the 20th century, economic growth accelerated. Electricity is the multiplier of economic growth. The 21st century is the century of gas. Industry, domestic heating and cooking, transport, power generation and some smelting uses gas. And what will the future centuries bring? Supplies of gas as methane hydrates are huge, cheaper methods of creating hydrogen are constantly being invented and we may finally enter a nuclear age with fission and perhaps fusion. For all of my life, nuclear fusion has been only 20 years away. It still is.

Energy in Australia

In Australia, electricity is generated from coal (73%), natural gas (13%), hydropower (7%), wind (4%), solar (2%) and bioenergy (1%)[597]. In 2013, Australia generated 247 TWh of electricity of which 234 TWh was sent to customers.[598] The difference of 15 TWh was used by the power stations themselves and 12.6 TWh was lost during

---

597: www.originenergy.com.au
598: http://www.world-nuclear.org/info/Country-Profiles/Countries-A-F/Appendices/Australia-s-Electricity/

transmission leaving 209 TWh for final consumption.[599] For Victoria, demand varies between 3,900 and 10,000 MW and in NSW between 5,800 to 15,000 MW.

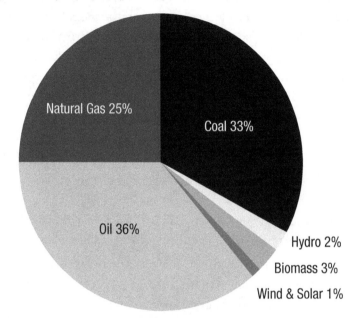

Figure 5: Australia's primary energy consumption for 2011[600]

Hot air comes from the green left environmental activists and misguided politicians about increasing the proportion of renewables. The large amounts of renewable energy that would be needed to keep people employed and homes on the grid can't be stored efficiently. The more renewable energy put into the grid, the more coal- or gas-fired back up would be needed. Rather than close coal-fired power stations, they would have to be kept running and/or replaced with modern coal-fired power stations as the electricity demand increases because wind power is unreliable and intermit-

599: http://www.world-nuclear.org/info/Country-Profiles/Countries175-A-F/ Appendices/Australia-s-Electricity/435
600: EIA International Energy Statistics

tent. The wind just does not decide to blow at times of peak power consumption.

In Australia, fossil fuels contributed to 86% of the total energy mix in 2011. Do the green left environmental activists have a better energy mix? How would food be delivered from farms to cities if fossil fuels were not used for heavy vehicles? Currently wind and solar in Australia are a very small amount of the primary energy consumption and the capital costs, recurrent costs[601] and environmental damage to expand this sector of the industry are horrendous.[602]

Australia is heavily dependent upon coal for electricity, three quarters of Australia's electricity is from coal-fired power generators. Historically, by world standards, Australia's electricity is low cost. This is why aluminium smelters were built in Tasmania, Victoria, NSW and Queensland. In 2008-09 about 11,000 kWh *per capita* was used, including that incorporated into exports (e.g. aluminium ingots).

Growth has since levelled out due to price rises, network costs and home solar generating systems. In 2013-14 rooftop solar panels resulted in 2.9% reduction in grid supply. Natural gas is increasingly used for electricity generation, especially in WA and SA. Electricity generation takes 44% of Australia's primary energy, and in terms of final energy consumption, electricity provides 24% of the total. There has been little new investment in base load electricity in Australia and new facilities will have to be built.

Australia is a large exporter of uranium, steaming coal and liquefied natural gas. This export of energy is used for generating electricity overseas. None of the uranium mined in Australia is used for generating electricity in Australia. Three times as much steaming coal is exported as is used in Australia and a significant amount of embedded energy is exported in smelted and refined metals such

601: Moran, A. 2014: Submission to the Renewable Energy Target Review Panel, 2014
602: Plimer, Ian 2014: *Not for Greens.* Connor Court

as aluminium (27 TWh/year i.e. 10% of Australia's gross electricity production) and zinc, lead, copper and nickel (14 TWh/yr). Over half the industry total of 80 TWh in 2012 was exported as embedded energy in refined metals. Most of the growth in value-adding manufacturing in the past 20 years has come from industries which are energy- and particularly electricity-intensive.

The past growth has occurred in Australia because of relatively low electricity prices coupled with high reliability of supply and the proximity of natural resources such as bauxite (NT, WA and Queensland), zinc ores (Macarthur River, N.T.; Century and Mount Isa, Qld; Broken Hill, and Cobar, NSW; Rosebery, Tas.), lead ores (Macarthur River, NT; Mount Isa and Century, Qld; Broken Hill and Cobar, NSW; Rosebery, Tas), copper ores (Mount Isa, Qld; Cobar and Parkes, NSW; Olympic Dam and Prominent Hill, SA; Rosebery, Tas) and nickel ores (Kambalda, Forrestania, Mount Keith, Leinster and Murrin Murrin, WA).

It is because Australia does not change governments with guns and has a stable democratic political administration that large long-term capital-intensive investments such as smelters and refineries can operate. This perception of stability is fast changing with a judiciary making political judgments on environmental and regulatory matters, a politicised bureaucracy, increasingly ineffective politicians, anarchistic unions and green left environmental activist movements.

A solution to fossil fuels

If the total fossil fuel electricity production is to decrease from 86% to 50% as Bill Shorten, the Federal Opposition Leader in Australia wants, then wind, solar and biomass will have to increase from 7% to 50%. Mr Shorten stated:

> *If we do not get serious about tackling climate change, if we don't get serious about investing in renewables, then we cannot say we are serious about economic reform.*

Neither Shorten nor anyone else have shown that human emissions drive global warming hence the need for "renewables" is not necessary. The "renewable" energy myth depends on the untruth that "renewable" energy reduces CO_2 emissions and this emitted CO_2 changes climate. Furthermore, the above statement demonstrates that economic reform is a joke. Investing in renewables actually means subsidising even more inefficient unreliable "renewable" energy.

When Paul Kelly (*The Australian*) asked:

> *How, as shadow treasurer, could you allow the leader to commit the party to the 50% renewable energy target without any framework of analysis, without work done on the impact of the scheme, the costs, any analysis whatsoever?*

The answer from the Labor shadow Federal Treasurer Chris Bowen was: "*Well, because I believe in renewable energy.*" This deflective throw away line is an admission that the whole global warming gravy train is a matter of religious faith. God help the taxpaying workers of Australia if Labor become a government and again enact what is considered popular policy made on the run. Sacrificing jobs and increasing costs to families for no climate gain is not good policy.

For a senior Australian politician to argue for an increase in renewables ignores some fundamentals. Australia's emissions amount to almost 1% of global emissions yet the Australian Labor Party wants renewable energy at twice the costs than that aimed by the world's largest emitters. The renewables mandate guarantees that the costliest capacity remains on stream whereas the cheapest capacity is prematurely scrapped. In 2014, renewables accounted for 98% of the 1,100 MW of added capacity.

By contrast, in 2014 coal has accounted for 90% of the 4,500 MW that has been shutdown or whose shutdown has been announced. With the renewables target nearly doubled as the Labor Party wants, the costs of electricity will rise by \$86 billion which will add another \$600 *per annum* to the average family on top of the already high electricity bill of \$1,600 *per annum* and will shutdown industry and drive even more jobs offshore. The 2014-2015 Australian deficit is less than half the \$86 billion that the renewable energy scheme will cost.

The cost of building wind and solar plants is horrendous, they damage the environment and they increase CO_2 emissions. Will increased taxation and electricity costs be used to pay for such white elephants or will the debt be increased to leave a problem for future generations.

It's all very well to claim that one is worried about the future environment that our grandchildren may inherit. What about the debt that they inherit from overspending for our glorious and wonderful ideological inefficient white elephants such as desalination plants, wind turbines and solar energy facilities? With 50% renewables in the mix, the end result will be to raise the price of electricity, to make electricity supply unreliable and to drive away employment-generating industry.

This hits the poor the hardest and some struggling employment-generating business will just close and move elsewhere. The policies of the political party that claims it is the workers' party will just increase unemployment. It's clear that the Federal Labor Opposition is looking at the populist vote in a forthcoming election, has not thought through the implications of Australia having 50% of its electricity from renewables and just does not care about the social, employment, personal and economic havoc their policy will create. Whatever it takes.

How would the green left environmental activists keep the lights on and keep people employed if there were a great reduction in the use of coal, oil and gas? Before subsidies can be given, there has to be a vibrant economy. No alternative energy company or investor is in the business to save the world. They are in the business to make money, something which is an anathema to the green left environmental activists.

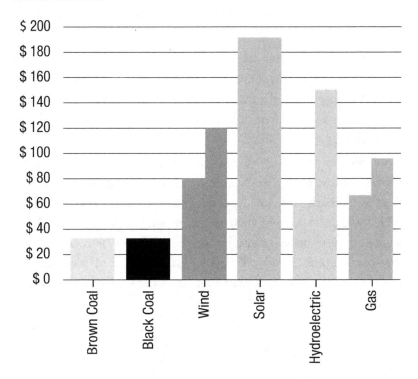

Figure 6: Costs of Australian energy per Megawatt hour (high and low).[603] **Electricity from renewable energy is far more expensive than electricity from coal.**

Hydroelectricity

New hydroelectricity plants could increase the proportion of renewables in Australia. The chances of the green left environmental activists and politicians allowing another hydroelectricity plant to be

603: Moran, A. 2014: Submission to the Renewable Energy Target Review Panel, IPA

built by damming rivers is zero, despite there being many favourable sites.

In low rainfall countries like Australia, it is only the coastal mountains that are a suitable site for hydroelectricity generation. In the north there are mountainous areas that receive high rainfall in the wet season. Elsewhere there are neither the rivers nor the topographic elevation difference needed for hydroelectricity.

Nuclear

If Australia wants to continue to grow and have large base load electricity for smelting, refining and general industry, then it must either continue with coal or go nuclear. A few hundred kilograms of uranium were produced in South Australia in the 1930s (Mount Painter and Radium Hill). Uranium was mined from 1954 to 1971 at Rum Jungle (NT), Mary Kathleen (Qld) and Radium Hill (SA), Mary Kathleen (Qld) (1976 to 1982) with Nabarlek (NT) (1979 to 1980), Ranger (NT) (1981 to present), Olympic Dam (SA)(1988 to present), Beverley (SA) (2000 to present) and Honeymoon (SA) (2011 to 2013).

There have been numerous other discoveries over the last few decades (e.g. Mulga Rock and Lake Way, WA). Australia has the largest resources of uranium in the world (31%) and is the third largest producer in the world (~8,000 tonnes contained uranium in yellow cake; 12% global supply) after Kazakhstan and Canada. Australia does not beneficiate yellow cake into fuel rods and does not have the capacity to reprocess used fuel rods. Some 35% of all energy exported from Australia is as yellow cake. Exports are to Europe (37.8%; mainly Belgium, Finland, France, Germany, Spain, Sweden and UK), USA (33.6%) and Asia (28.6%; mainly Japan, South Korea, China and Taiwan). USA generates 30% of global nuclear power.

There are 30 countries that have 438 nuclear reactors generating electricity. Another 67 reactors are under construction in 15 countries.[604] If emissions of CO_2 were really a problem for the green left environmental activists in Australia, then they would welcome a long-term program of building nuclear reactors in order to greatly reduce Australia's CO_2 emissions. The long term construction jobs and highly-skilled workers required to run reactors would create jobs for generations, as would the additional electricity.

If one believes that CO_2 emissions from coal-fired power stations are harmful or carbon emissions taxation becomes expensive, then nuclear power is the least-cost low-emission tried-and-proven technology that can provide the needed base load electricity.

Australia is a dry country, some 400 Giga litres of water is lost from evaporation in the cooling towers of coal-fired power stations hence seawater would be the best fluid for cooling nuclear reactors thereby releasing fresh water for other purposes.

Australia is well blessed with uranium ores and an uninhabited hinterland if reprocessing needs to be isolated. However, acts of Federal and state parliaments would have to be revoked if Australia was to go nuclear.[605]

Between 1958 and 2007, the 10 MW High Flux Australian Reactor (HIFAR) operated at Lucas Heights (NSW) and was used for materials research, production of medical isotopes and for irradiation of silicon for the computer industry. In 2006, it was replaced by the 20 MW (Open Pool Australian Light water reactor (OPAL) which is in the top league of the world's 240 research reactors. This reactor will soon produce 25% of one of the world's medical isotopes.[606]

604: http://www.nei.org

605: Federal Acts: ARPANS Act 1998, Section 10, Prohibition on certain nuclear installations and the EPBC Act 1999, Section 140A. State Acts: Victoria Nuclear Facilities (Prohibitions) Act 1986, Queensland Nuclear Facilities Prohibition Act 2006, NSW Uranium Mining and Nuclear Facilities (Prohibition) Act 1986.

606: Mo^{99} precursor to Tc^{99}

If the Pope really wanted to lower emissions then he would be agitating for a global nuclear industry.

Environmental impact of solar power

The influence of Ra

Throughout most of human history, there has been Sun worship. We have known for a very long time that without the Sun there is no life on Earth.

Some 2,000 years ago, Archimedes dreamed that solar energy could be harnessed. Dreamers, Sun worshippers, hippies, the unbalanced and politicians seeking the green left environmental activist vote think that solar will provide the electricity needed for a modern industrial society. Solar cells (photovoltaics) were invented in 1839. One would have thought that if solar cells were to be a low-cost efficient competitive energy dense system to assist humanity, then 175 years of refinements and improvements would have been enough to make it cheap and efficient. Apparently not. After all, this time period was enough to make the steam engine efficient and we still use steam today for the generation of electricity from coal and nuclear fission. During this time, the internal combustion engine was invented and, for the last 120 years, has undergone significant improvements such that hyper-efficiency is gained by modern small engine capacity turbo-charged diesel engines.

We are told that solar power is clean, free, renewable and will go on forever. This is true until one looks at the uncomfortable fundamentals such as the environmental and monetary costs. Solar power exists only because of subsidies and the mistaken belief that it reduces CO_2 emissions. It does not. Solar power, like wind power, requires infrastructure construction that actually adds to human emissions of CO_2.

Solar cells

In 1839, a silicon solar cell was 10% efficient and this was excluding light reflection and current leakage. It still is. Why? A silicon solar cell does not create energy, it converts just one wavelength of the whole spectrum of solar energy into electrical energy and the rest of the spectrum creates heat energy. Just one wavelength in the infrared spectrum (1,130 nm) excites an electron to jump to a higher energy state, when the electron jumps back it gives out a small amount of electricity (1.1 eV).

Solar cells do not create energy, unless the laws of thermodynamics can be changed because of ideology. They convert light into electricity. This is why even the best silicon solar cells have an efficiency of not much higher than 10% and why we shouldn't wait around expecting huge efficiency improvements. In reality, efficiency is even lower because of light reflection and current leakage.

Technological improvements

If green left environmental activists want solar energy to be more efficient, then they had better invent a few new laws of physics and persuade electrons to respond to a great spectrum of wavelength rather than one specific wavelength in the infrared part of the spectrum. Don't wait up. The only reason Western countries have solar electricity is that it is subsidised. In remote areas with small power needs and where regular maintenance is prohibitively expensive, solar power is used for lighting, telecommunications, navigation beacons, recording equipment, marine buoys, electric fences, pumps at bores and satellites. This is the market talking. Farmers I talk to also tell me that solar panels for fences, dams and bores only last about five years before they need to be replaced.

The US Department of Energy concluded that solar electric systems couldn't meet the energy demands of an urban community or industry. Large scale solar arrays, whether photovoltaic or solar thermo-electric are far too variable, unreliable, expensive and ecologically damaging. Solar can only make a minor contribution to the national power requirements. Solar power is not very efficient and the optimal figure used for incident radiation is 10 watts per square metre with an overall system efficiency of 5%.

Other factors affect the efficiency of solar radiation such as latitude, time of year, time of day, light reflection, current leakage and aerosols. There are also long-term weather fluctuations due to cycles of cloud coverage that can change the efficiency by up to 4%. Aerosols can reduce the efficiency by almost 30%. Furthermore, in remote areas, lack of regular cleaning off of dust and plant spores from the glass surface covering the photovoltaic cells can result in reductions of efficiency by up to 50%.

Some green left environmental activists argue that we should invest in solar power as the breakthroughs are just around the corner. What breakthroughs? There are still no cost-effective batteries to store solar energy for use at night and we have been waiting for such breakthroughs for a century. The only sensible suggestion is to use solar energy to pump water uphill into dams and generate electricity when needed at peak times or at night. Dams are off the menu for green left environmentalists.

Surely new developments with metals, metalloids and super conductors are just around the corner and these can be used to make more efficient solar cells? Well … yes and no. Cells of higher efficiency require the use of exotic, rare and poisonous elements such as germanium, gallium, indium and cadmium. There are no germanium, gallium, indium and cadmium mines in the world and these metals are by-products from the zinc, aluminium and tin smelting and refining industries.

To produce 1% of the US electricity requirements from a germanium or gallium solar cell would require three times the planet's annual production of germanium and twenty times the world's annual production of gallium. The zinc (for germanium) and aluminium deposits (for gallium) are yet to be discovered and, if they were, it would not be economic to produce massive excess amounts of zinc and aluminium just to provide small amounts of germanium and gallium. Zinc and aluminium are the two metals with the highest amounts of embedded energy and, to create small volumes of cheap solar power using other materials, astronomical amounts of conventional base load energy would be required.

For every kilogram of aluminium made, 15 kWh of electricity is needed. For example, some 10% of all of Australia's electricity is embedded in aluminium (27 TWh[607]) made in Australia and exported. Non-ferrous metals smelted and refined in Australia use 43 TWh of the 80 TWh used in Australian industry in 2012. If Australia was to become a germanium and gallium producer, then it would need a far greater generating capacity and electricity would have to be reliable and cheap. This just cannot be provided by the solar or wind industries.

If more gallium were to be produced for solar cells, the already marginal aluminium industry would have to greatly increase production, flood the market with massive quantities of unwanted aluminium and use stupendous amounts of electricity. Furthermore, these elements are far more expensive to produce than silicon (the second most abundant element on Earth). In the US, production of these germanium or gallium solar cells would require 17% of the US annual cement production. To make cement, limestone needs to be burned and CO_2 is released to the atmosphere.

Energy efficiency

Efficiency is a banned word from the green left environmental activist lexicon. However, as few of these folk read books I can safely continue. If a solar panel is to generate maximum electricity, condi-

---

607 1 TWh = 1 billion kWh

tions must be ideal (i.e. middle of the day, clear sky, low latitude). The maximum incident solar radiation value is 1,000 watts per square metre. It is claimed that an off-the-shelf solar panel will produce 110 watts for a one metre square panel. This is an efficiency of only 11%.

Solar panels can be flat or can use concentrating collectors and solar trackers to give maximum incident solar radiation for a longer time. In high winds, these concentrating collectors and trackers have to be closed down. Long-term measurements show that the average incident solar radiation is 125 to 375 watts per square metre.

Assuming a very generous optimistic efficiency for the average solar panel of 15%, an average of 19 to 56 watts per square metre would provide only 0.46 to 1.35 kilowatt hours per metre per day at an average of 0.61 kilowatt hours per square metre per day. The glass cover that protects the solar cells reduces the efficiency to 13% and further system losses of 7% can be expected due to localised conditions and the conversion of direct current (DC) to alternating current (AC) that we use in our domestic life.

The US Department of Energy calculated that solar panels have a 10.27% efficiency that equates to 4.25 kilowatt hours per square metre output per day. Solar power advocates quote the maximum radiation value of 1,000 watts per square metre and don't really worry about those trivial little inefficiencies. But then again, the world of the green left environmental activist is all about inefficiency with costs passed on to the consumer.

Solar power has a low capacity factor. For example, in Germany, it is about 10%. Hence 10,000 megawatts of solar power capacity is needed to generate the same amount of electricity as a 1,000-megawatt thermal coal or nuclear power station. Furthermore, when the 10,000-megawatt solar power generator is producing its maximum of 10,000 megawatts, the grid system cannot cope and hence huge yet-to-be-invented energy storage systems are needed or the solar power station needs to be shut down. Now that's efficiency!

If green left activists really want efficient solar energy, then the toxins germanium, gallium, indium and cadmium could be used. They would, of course, be aware that there are no germanium, gallium, indium and cadmium mines in the world, that these elements occur as trace elements in zinc, aluminium and tin ores and are extracted during CO_2-emitting smelting processes. Zinc, aluminium and tin have a huge amount of embedded energy and would have to be produced on an unprecedented scale resulting in a massive use of energy.

Furthermore, to produce 1% of the US electricity requirements from a germanium or gallium solar cell would require three times the planet's annual production of germanium and twenty times the annual production of gallium. In order for the green left environmental activists to have more efficient solar cells, there would have to be massive mining of toxins on a scale never before seen on Earth. And for what purpose? To produce an inefficient solar panel.

Is this what the Pope wants? Free markets are highly competitive and if there was a better way of creating solar panel, a competitor would fill the gap. But then again, the Pope seems to be very strongly against free markets, market forces and capitalism,[608] all of which have made the world a far better place and contrast with failed centrally managed socialist systems. The Pope offers no solution to global energy problems.

Costs

Costs, efficiency and productivity are dirty words for the green left environmental activists so let's just whisper dirty for a while.

608: *Laudato Si'*, Paragraphs 30, 51, 55, 56, 94, 109, 123, 129, 190, 190, 195, 203, 209, 210 and 215

One square metre of a solar panel costs $750. Installation doubles this cost. For a 1,000-megawatt plant to provide electricity in the depths of winter, 3,230,000 panels are required at a bargain basement price of $4.83 billion. This is only for peak production of 1,000 megawatts at the optimal time of day. If the solar power station were to compete with a conventional coal-fired thermal power station providing 1,000 megawatts constantly with a load factor of 70%, capital costs for a solar power station would be in the order of $100 billion. The latest 1,000 megawatt coal-fired thermal power station built in Australia cost $1 billion at current costs and over its 20 year life would consume $2 billion worth of coal. This, of course, assumes that a solar power station would last 20 years at peak efficiency. An optimist would give it five years, at best 10 years because wafers of silicon quickly reconstitute and become even less efficient.

There are claims that solar panels are becoming cheaper and cheaper. This may be so but is does not change the facts. Solar power is still far too expensive, too unreliable and too environmentally damaging. Whatever the cost of panels, solar power cannot compete without massive subsidies. A recent study in Germany showed that solar power is four times as expensive as power from a prototype nuclear reactor being built in Finland, which is a particularly expensive design.

When wind and solar electricity can so easily be shown to be uneconomic, why on Earth should there be a "renewable" energy target? When it is obvious that CO_2 does not drive global warming, why should there be costs and restrictions on employment-producing industries that emit plant food?

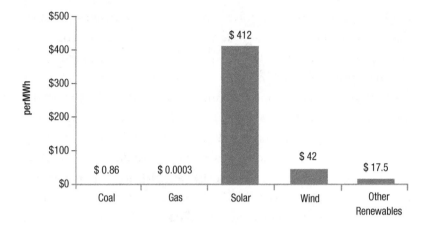

Figure 7: Government subsidies for electricity production in Australia by source, 2013-2014.[609]

From any perspective solar power is too expensive, too environmentally damaging and cannot provide large-scale energy to an electrical grid system. The greens hope against all knowledge that a large-scale low cost method of storing electricity for days, months or years is just around the corner. Such technology does not exist and is not even on the horizon. How could a Third World country afford inefficient unreliable costly solar power?

In the UK, solar panels generate barely 1% of all electricity at only 10% of their 6.5 GW rated capacity. On 11 April 2015, when solar-generated electricity rose from zero to 3.7 GW, before falling back again to nothing within a few hours, the National Grid had to pay out £500,000. Most of this was to compensate wind facilities for switching off 2.5 GW of wind-generated power. And who really paid? The consumer. Early closures of solar subsidies would save between £40 and £100 million in 2020-2021 in the UK.[610]

609: http://www.minerals.org.au/file_upload/files/media_releases/Electricity_production_subsidies_in_Australia_FINAL.pdf

610: https://www.goc.uk.government/upload/systems/upload/attachment_data/file/447323/Solar-PV-within-the-RO-consultation_-_Impact-Assessment.pdf

The normal suspects such as Friends of the Earth and subsidised solar providers such as Good Energy are squealing. During winter in Western countries, many unnecessary deaths of older people are caused, or hastened, by fuel poverty. Far more people die in winter from the cold than in summer from the heat. Fuel poverty has been worsened by environmental policies resulting in massive increases in energy costs in order to subsidise unreliable, uneconomic and intermittent electricity sources such as wind and solar.

The UK Department of Energy and Climate Change shows frightening statistics on fuel poverty in the UK.[611] A 2011 YouGov poll showed that up to 24% of UK households suffer fuel poverty and a Confused.com survey showed that 82% of the UK population are concerned that they may not be able to pay their energy bills during winter. This just should not happen in a developed country and is the end result of poor energy policy, the same policy that the Pope advocates. Anti-fossil fuel activists have blood on their hands. There is no virtue in supporting renewable energy when it makes people poorer and hastens the demise of the weakest in society.

If anyone is not sickened by this sort of behaviour, then they are morally bankrupt. The Pope, a leading figure on morality, should have also been sickened and written about this in his Encyclical. I guess his green left environmental activist advisors didn't seem to think morality was important.

Environmental devastation

A 1,000 MW nuclear or coal-fired power station occupies an area of 30 to 60 hectares (75 to 150 acres). A 1,000 MW solar power station would have to produce enough energy for an 8-hour day plus reduced energy production for the remaining 16 hours. The area required is 55.5 square kilometres. To do this with an efficiency of

611: https://www.gov.uk/government/collections/fuel-poverty-statistics

10.27%, the area of solar panels, space between panel to prevent shading and maintenance roads would have to be 128 square kilometres (50 square miles; 12,800 hectares). All plants (and hence animals) would be removed from this 128 square kilometre area just to produce ideological inefficient unreliable electricity. This is green left environmental activism at its very best.

To build a 1,000-megawatt solar power station, it is not only the solar cells that are needed, there are structural supporting materials, concrete foundations, transmission systems, access roadways and a dispersed area of collectors and AC converters. The amount of material required is huge. Furthermore, massive earth works would have to be undertaken by diesel machinery emitting CO_2 during site preparation.

To produce the 35,000 tonnes of aluminium for structural support 12,777,950,000 kWh of electricity is needed which is now embedded energy and 735,000 tonnes of CO_2 is released during the process of smelting and refining to produce the 35,000 tonnes of aluminium. Some 75,000 tonnes of glass is required to cover solar panels and the manufacture of this glass releases 260,000 tonnes of CO_2 into the atmosphere. The embedded energy in the glass is 661,425,000 kWh of electricity.

At least 600,000 tonnes of steel are required for the 1,000 MW solar industrial plant and electricity used for the manufacture of this steel would release 3,901,576 tonnes of CO_2 and the blast furnace would release 1,218,000 tonnes of CO_2 into the atmosphere. The embedded energy in the steel is 9,070,000,000 kWh of electricity. For wiring, some 7,500 tonnes of copper would be required. It has embedded energy of 529,507 kWh. To make this copper, 13,500 tonnes of CO_2 would be released from the smelter and the electricity used to run the smelter would release 494,200 tonnes of CO_2 into the atmosphere.

The two million tonnes of concrete for footings would release 2,706,885 tonnes of CO_2 for the electricity used and 360,000 tonnes of CO_2 from burning limestone for cement manufacture. The embedded energy in the concrete is 6,249,000,000 kWh of electricity. Just for these components alone, some 9,688,661 tonnes of CO_2 would have to be released to the atmosphere and the 1,000 MW generator would have to work at an efficiency of 10.27% for over 24 years just to pay back the 20,804,705 kWh of embedded energy.

Unless solar power generation is very heavily subsidised, it is clearly uneconomic. Why should the average worker pay for environmentally devastating uneconomic subsidised electricity?

These are minimum figures because CO_2 emissions from road building, road and shipping transport, site machinery, manufacture of other metals (e.g. silicon in solar cell, tin/silver at electrical contacts), packaging, office activities etc have not been calculated. Nor have the use of vehicles for maintenance. Over 100,000 truck-loads of concrete would have been delivered by diesel-powered trucks emitting CO_2 and particulates and creating dust.

There are numerous poisonous, flammable and hazardous chemicals used in the manufacture of a silicon solar panel. For example, arsenic, cadmium and lead are used in solders and a cleaning fluid for silicon manufacture is sulphur hexafluoride, a greenhouse gas that is 25,000 times more powerful than CO_2. A recent calculation showed that to manufacture a 2 gram silicon chip, 72 grams of chemicals, 1.6 kilograms of oil or coal equivalent and 3.2 tonnes of water are used.

The real environmental cost for the manufacture and decommissioning of solar cells is not known but it is not pretty. And all this for a short-life solar cell. When solar power was all the rage, numerous Chinese companies were established to manufacture and sell solar cells to credulous Western green-contaminated countries. These companies are now disappearing at a very rapid rate, as are the Chinese poisoned by the pollutants used to make solar cells. Presumably for the greens, saving the planet is worth the human cost of killing Chinese workers.

The use of huge amounts of energy and the release of at least 10 million tonnes of CO_2 into the atmosphere just to construct a 1,000 MW solar powered generator in order to save the planet from increased CO_2 into the atmosphere does not look like a good idea, especially as at least 128 square kilometres of plant and animal habitats would be destroyed and the energy produced would be inefficient, unreliable and costly. Furthermore, because nature does not co-operate with dreaming ideologists, coal- or gas-fired electricity would have to be used as backup and would keep emitting the CO_2 that was meant to be saved and not emitted to the atmosphere.

Solar power generation at night

Sunny Spain was touted as the perfect place for solar power generation. Spain spent a fortune on constructing solar and wind power generators with generous subsidies. Spain became so clever at generating solar electricity that it even managed to do it at night. Generating solar electricity at night? No, it is not the new physics.

It is because subsidies were so incredibly high that solar power companies could make money by illuminating solar panels with floodlights at night. The floodlights were powered by diesel generators. This is madness. No wonder Spain went broke.

Subsidised hot air and sunshine

The Premier State

It has become political fashion to promote wind and other renewables. But there are some realities. In 2015, NSW had no solar power generation capacity whatsoever and, although there are five wind in-dustrial plants, the total wind power input into the grid was 0.6%. And at what cost? No green left environmental activist in NSW could possibly survive day-to-day life on wind or solar power from the grid.

Even green left environmental activists need fossil fuels to trans-port food from rural areas or abroad to the cities. A huge amount of food, flowers and high unit value materials flit around the world as air cargo. Without fossil fuels and dams for domestic and industrial electricity in NSW, there would be no cooling, heating, cooking, re-frigeration, transport, employment or communications. Nothing. If that's the world that the green left environmental activists want us to live in, then let's see them lead by example and shout their ideology from the caves on a cold, wet and windy night.

In NSW in 2014, there was a mix of electricity generation with coal, natural gas, hydro, wind, diesel and coal seam gas.[612] The table shows what we already know. If heavy-duty base load power is need-ed, then coal (10,760 MW; 59.3%) and hydro (4,510 MW; 24.9%) do the heavy lifting 24/7. Gas (2,144 MW; 11.8%) and hydro can be used to increase electricity generation in peak load times. Of the 18,138 MW capacity, wind capacity is 550 MW (3.0%) but, because the wind does not blow all the time, the input into the grid is about 110 MW (0.6%). The sugar cane waste (bagasse) only has a 68 MW (0.4%) capacity and electricity generation is seasonal. Diesel (106 MW; 0.6%) is used when required.

More than 71% of electricity in NSW is generated from fossil fuels. In Australia, 65% of Australia's electricity needs are produced from 48% capacity reflecting the predominance of base load de-

---

612: http://www.aemo.com.au/About-the-industry/Registration/Current-Reg-istration-and-Exemption-lists2465

mand and the fact that coal provides the main base load demand in
Australia.

NSW is taking a great step (and more than likely economically
backwards) in building a solar power plant at Broken Hill. At capacity, the solar plant will provide a glorious 2.9% or at normal capacity
0.29% of electricity for NSW. The 53 MW capacity plant is privately
owned and the capital cost is $166.7 million. The NSW government
and Federal government's Australian Renewable Energy Agency (i.e.
taxpayers) are providing $64.9 million of the capital cost[613]. Further-
more consumers will be forced to pay extra for this "renewable"
energy.

Given the well-known inefficiencies of solar plants, the reality is
that about 5-6 MW will be available from this new solar plant. This is
not enough to keep the mines operating at Broken Hill. And if there
are no mines, then there are no jobs. I often go underground in the
zinc-lead-silver mines at Broken Hill. Do I rely on solar-generated
electricity when underground at night to keep the safety systems and
pumps operating? Will the processing plant be able to operate 24/7
on solar energy?

Whatever the green left environmental activists may desire, the
reality is that NSW will continue to have a reliable coal-hydro-gas
mix for peak- and base-load electricity. Why? Because it works.

613: http://www.agl.com.au/about_agl/how-we-source-energy/renewable-ener-
gy/broken-hill-solar-plant

Table 2: Major existing power stations in NSW[614]

Power Station	Location	Technology	Capacity (MW)
Appin	Illawarra	Gas[a]	55
Bayswater	Hunter	Coal	2,800
Bendeela	Nowra	Hydro	240
Blowering	Snowy	Hydro	80
Broadwater	North Coast	Sugar cane waste	38
Boco Rock	Nimmitabel	Wind	113
Broken Hill	Broken Hill	Diesel	50
Capital	Tarago	Wind	140
Condong	North Coast	Sugar cane waste	30
Colongra	Central Coast	Gas[b]	724
Cullerin	Upper Lachlan	Wind	30
Eraring	Hunter	Coal	3,000
Eraring	Hunter	Diesel	56
Gullen Range	Goulburn	Wind	172
Gunning	Gunning	Wind	47
Guthega	Snowy	Hydro	80
Hume	Snowy	Hydro	70
Hunter	Hunter	Gas[b]	50
Liddell	Hunter	Coal	2,200
Mt Piper	Central West	Coal	1,400
Murray	Snowy	Hydro	1,575
Smithfield	Smithfield	Gas[c]	175
Tallawarra	Illawarra	Gas[c]	435
Tower Point	Illawarra	Gas[a]	41
Tumut	Snowy	Hydro	2,465
Uranquinty	Wagga Wagga	Gas[c]	664
Vales Point	Central Coast	Coal	1,360
Woodlawn	Tarago	Wind	48

a = coal seam methane; b = open cycle gas turbine; c = combined cycle gas turbine

614: http://www.resourcesandenergy.nsw.gov.au/investors/projects-in-nsw/electricity-generation

In 2007, Australia had one of the lowest electricity costs in the world. Between 2007 and 2013, Australia had a Labor Federal government who brought in a carbon tax and various federally funded renewable energy schemes. In 2011, the electricity prices in Western Australia, Victoria, NSW and South Australia were so high that they ranked just behind Denmark and Germany.

Farewell jobs

A 2009 testimony about Spanish renewable energy to the US House Select Committee on Energy Independence and Global Warming,[615] showed that for every green energy job financed by Spanish taxpayers, 2.2 jobs were lost. Only one out of ten green jobs were in maintenance and operation of already installed "alternative" energy plants and the rest of the jobs were only possible because of high subsidies. Each green job in Spain cost the taxpayer $750,000 and green programs led to the destruction of 110,500 jobs. Each green energy megawatt installed destroyed 5.39 jobs elsewhere in the economy. I'm sure those pushed into unemployment by green activism feel that they have made a sacrifice for a higher cause.

On 16 January 2009, during a visit to an Ohio wind turbine component manufacturing business President Obama stated:

> *And think of what's happening in countries like Spain, Germany and Japan, they're making real investments in renewable energy. They're surging ahead of us, poised to take the lead in these new industries.*

President Obama is doing a pretty good job of increasing unemployment. If he follows the Spanish example he can do even better. Spain has since gone bankrupt, part of which was due to the extraordinarily high cost of electricity and subsidies. Thanks to the greens. And the green left activists feel smug and take the moral high ground because of their policies.

615: http://www.markey.senate.gov/GlobalWarming/index.html

Environmental impact of wind power

The answer is not blowing in the wind

The narrative is that by burning coal to generate steam to drive a spinning magnet to produce cheap electricity, the CO_2 emissions from coal burning will raise atmospheric temperature which will then create a sea level rise because of water expansion and ice melting. Therefore, to avoid this predicted global catastrophe we must build industrial sites with wind turbines. This is the green left environmental activist narrative. However, as CO_2 does not drive global warming, building wind industrial facilities is not necessary. Furthermore, wind energy is costly, inefficient and damages the environment.

Miguel de Cervantes' *Don Quixote* saw windmills as evil giants fit only for destruction. I do also but for different reasons from the Don. Too often when the green left has lost arguments of science (on the basis of weak evidence), economics and logic, they invoke the "precautionary principle". There is no such principle in science. The Pope quotes from the Rio Declaration of 1992[616] the statement wherein we first see the so-called precautionary principle:

> ... *where there are threats of serious irreversible damage, lack of full scientific certainty shall not be used as a pretext for postponing cost-effective measures.*

The Pope's precautionary principle

The Pope indirectly acknowledges scientific uncertainty in his Encyclical, promotes alternative energy technologies and yet fails to address the painful well known facts about wind energy. Why doesn't the Pope use the precautionary principle regarding the health effects of wind turbine noise? Low frequency noise, possibly as low as 8 Hz,[617] has possible physical and thence psychological effects on hu-

---

616: *Laudato Si'*, Paragraph 167
617: Physikalisch Technische Bundesanstalt EurekAlert Public release 10-Jul-2015; Dr Christian Koch

mans and it is not known what effects low frequency noise has on animals. However, mammals feel distress, confusion and fear from the unheard infrasound in the roar of the big cats.[618] The Pope's advisors should have been very cautious about the health effect of wind turbines when promoting alternative energy for humans. These medical effects of noise from wind turbines at this stage are difficult to quantify, medical research is at an early stage, further work is needed but reports from residents living near wind turbines suggest serious unresolved problems resulting from low frequency infrasound.[619]

The Pope's green left environmental advisors are only too aware of or could have found out the effects wind turbines have on human health[620] and, if the Pope did not know, then his advisors should have told him. His Holiness should have been invoking the "precautionary principle" for wind energy, an alternative energy championed by many green left activists. He didn't. Why not? Green left activists are very vocal about what can't be seen (e.g. radiation, GM crops) but are hypocritically silent about what can't be heard. Inaudible sound maybe can't be heard but it can be felt.

There is extensive literature to show that wind turbines kill birds (especially rare eagles), bats and are built in areas of outstanding natural beauty. The Pope shows concerns for landscapes[621] and especially rural landscapes[622] and, because wind industrial developments are in rural landscapes, it is surprising that the Encyclical did not mention the appalling environmental damage that wind turbines create in the name of environmentalism. The Pope states in Paragraph 151:

618: von Muggenthaler, E. 2000: The secret of a tiger's roar. *American Institute of Physics*, www.sciencedaily.com/releases/2000/12/001201152406.htm

619: Australian Senate Select Committee on Wind Turbines, 2015

620: Salt, A. N. and Hullar, T. E. 2010: Responses of the ear to low frequency sounds, infrasound and wind turbines. *Hearing Research* 286, 12-21

621: *Laudato Si'*, Paragraphs 21, 58, 184 and 232

622: *Laudato Si'*, Paragraph 151

Interventions which affect the urban or rural landscape should take into account how various elements combine to form the whole which is perceived by its inhabitants as a coherent and meaningful framework for their lives.

No mention of wind turbines as an environmentally damaging "renewable energy". In many places in the Encyclical, the Pope is an enthusiastic cheer-leader for renewable energy and clearly has not been informed of the details.[623]

Environmental damage

The theory is that wind-generated electricity decreases emissions of CO_2. This does not happen. They increase emissions. Wind turbines produce intermittent and unreliable electricity hence require 24/7 backup by a coal-fired power station which cannot be just turned off and on. Wind generating facilities cannot exist without fossil fuels. The energy density of wind is very low. If all of the electricity requirements of the USA were to be from wind, then an area the size of Italy would be required.[624] The resources that a wind turbine uses just don't seem to get mentioned by the green left promoters.[625]

In addition, the energy required to make the steel column, blades, concrete footings, copper wiring and rare earth magnets is more than the wind turbine will ever produce in its working life.[626] There is a long history of dreadful pollution associated with the production of rare earth elements for wind turbine magnets, especially in China, and yet these metals are a fundamental component of wind turbines. Additional energy from fossil fuels is expended with transport of

623: *Laudato Si'*, Paragraphs 26, 52, 153, 164, 165 and 179
624: Bryce, Robert 2014: *Smaller, faster, lighter, denser, cheaper. How innovation keeps providing the catastrophists wrong.* Public Affairs
625: Wind energy in the United States and materials required for the land-based turbine industry from 2010 through 2013: US Geological Survey
626: Plimer, Ian, 2014: *Not for greens.* Connor Court

great tonnages of concrete, maintenance and decommissioning. However, many wind industrial complexes are not decommissioned and are left as rusting, commonly burnt out, relics in areas of great scenic beauty.

The land area used by wind industrial complexes is orders of magnitude greater than the land area used for gas generators, coal-fired power stations and nuclear power stations. A single unsubsidised shale gas pad of 2 hectares would produce as much energy in 25 years as 87 giant wind turbines covering 15 square kilometres (with turbines visible from 30 kilometres away).[627] Furthermore, shale gas used to drive turbines to make electricity is not intermittent and unreliable.

The end result of environmentalism as demonstrated by wind industrial complexes is damage to the environment. Environmentalism tries to make everyone feel guilty about the environment. This guilt is a hangover from Western Christianity and the papal Encyclical fuels this fire.

An each way bet

A few simple numbers should have sobered the Pope's enthusiasm for renewable energy. If the wind turbine is actually working, then one megawatt of wind power requires 103 tonnes of stainless steel, 402 tonnes of concrete, 6.8 tonnes of fibreglass, 3 tonnes of copper and 20 tonnes of cast iron. Stainless steel requires a long chain of processes[628] of mining, smelting and fabrication. All require fossil fuels. Mining involves high-density energy such as diesel fuel. About 2 litres of fuel are used for every tonne of rock produced.

Transporting iron ore from the mine to the steel mill requires diesel and ship bunker fuel. Many ship fuels are dirty and emit large

627: Booker, Christopher 2015: Why are the greens so keen to destroy the world's wildlife? *Daily Telegraph*, 4 July 2015
628: Plimer, Ian 2014: *Not for Greens*. Connor Court

quantities of sulphurous gases into the atmosphere. To convert iron ore into steel, coking coal or rarely natural gas is used and CO_2 is vented into the atmosphere. The fossil fuels for steel manufacture are for both energy and chemical reduction of an oxide to a metal. Chemical reduction cannot be done with wind, solar, hydro or nuclear power.

Concrete is composed of aggregate, sand and cement. The precursor to cement is limestone and shale, these are heated and CO_2 is vented to the atmosphere. To quarry, transport and crush gravel, sand, limestone and shale requires diesel fuel. The energy to heat limestone to convert it to cement requires coal or natural gas.

It doesn't end there. Large wind turbines need to extract energy from the grid to start and when the turbine is not spinning it still requires energy for the controls, lights, communications, sensors, metering, data collection, oil heating, pumps, coolers and gearbox filtering systems. This comes from the grid and is provided by burning coal. The bottom line is that wind turbines cannot be built, operated or maintained without using fossil fuels.

Coal-fired power stations are not shut down because wind turbines provide electricity. Coal-fired power stations must be kept running for when the wind does not blow. A wind turbine that produces inefficient and intermittent expensive subsidised electricity results in the emission of more CO_2 than if the same amount of electricity was generated from a coal-fired power station. And, of course, there is the killing of birds and bats and damage to human health.

None of this was mentioned by the Pope. He was poorly advised. The Pope objects to the industrial processes that build a renewable energy facility such as a wind turbine because of energy use and waste creation.[629]

The Pope can't have an each way bet.

629: *Laudato Si'*, Paragraphs 21 and 161

Global hot air day

Global Wind Day is 15 June. This is a feel good orchestrated media event celebrated by rent seekers with their snouts in the trough getting subsidies and raising electricity prices in a guaranteed market, landowners receiving compensation indirectly from the consumers and the green left environmental advocates promoting various UN agendas at the expense of sovereignty, freedom, the environment and financial common sense.

Global Wind Day is not celebrated by birds, bats, nearby residents who suffer decreased property values, humans proximal to wind turbines suffering health problems from thumping blades and consumers and industry hurting by paying higher electricity prices. Although aesthetics is subjective, most folk see no scenic enhancement in hilltops covered by wind turbines. Electricity consumers, taxpayers, businesses and true environmentalists only hear hot air on Global Wind Day. Maybe there are not enough days in the year to set aside special days to celebrate employment destruction, industry destruction, fuel poverty, sovereignty loss, freedom loss, wealth transfer and UN hypocrisy.

A failed concept

Wind power reached its zenith 400 years ago. Centuries ago, wind was used for pumping water and grinding grain and these processes did not have to be undertaken all the time. Since then, the increasing energy requirements, energy density, inefficiency and unreliability made wind power more and more expensive. Wind-generated electricity is massively subsidised. Despite subsidies for competing energy such as wind and solar, coal producers and coal-fired electricity generators are not subsidised and face punitive regulatory barriers and massive environmental and legal costs just to establish an employment-creating business. It is only by mandating and subsidising increasing proportions of electricity consumption from the highest cost "renewable" sources, thereby driving up electricity prices, that governments have made cheap coal-fired energy more expensive and comparable in price to wind energy.

Wind industrial complexes to generate electricity are a failed concept. It's time for the supporters to pay for their fantasies instead of forcing taxpayers and power consumers to write the cheques. The wind industry has been promising to deliver competitively-priced electricity but has never achieved it. Australia has a sad history of funding infant industries and subsidising industry (e.g. textiles, clothing, shoes, cars, chemicals and now wind energy). They all failed. A small country like Australia can not back losers and subsidise high capital cost technology even if well funded lobby groups frighten the public, pressurise politicians and have a supportive media who have not the skills to ask searching questions. Such support only raises regulatory costs, damages the economy, jobs and the budget and increases debt. The Australian government's wasteful green bank, the Clean Energy Finance Corporation, receives subsidies from consumers through the Renewable Energy Target. The so-called green energy is twice as expensive as that from more reliable coal-fired power stations and the annual subsidy is $2.5 billion.

Wind is not free

Wind is not free. The green left environmental activists can't do sums. Wind power is horrendously expensive, damages the environment and is unreliable. The US Senate Finance Committee[630] has proposed a 2.3 c/kWh production tax credit for wind energy. This is the seventh time since 1992 that a subsidy has been granted to help "the industry compete in the marketplace" and there have been other "temporary" federal subsidies since 1978. One would have thought that after more than 30 years, wind power would have been able to

630: http://finance.senate.gov

compete. Unless, of course the industry is so hopeless that it cannot survive without massive subsidies. The Energy Information Administration[631] stated that the 2013 production tax credit for wind was $US5.9 billion and solar $US5.3 billion. In 2013, wind and solar in the US provided less than 5% of the total electricity and yet received 50 times more subsidy than coal and gas combined.[632] These additional costs are borne by the taxpayer, thanks to renewable energy mandates in 29 states and the District of Columbia that are designed to guarantee a market share no matter what the production costs for wind and solar might be.

For example, households in New York State[633] now pay $US400 *per annum* more than the national average for electricity[634]. Statewide this 53% extra cost over the national average is $3.2 billion a year. The 15 wind industrial complexes operating in 2010 produced an output of 2.4 million MW hours. The same amount of electricity could be generated by a small 450 MW gas-fired generating plant operating at 60% capacity and with a capital cost of 25% of the wind turbines. The wind turbines need replacing every 10-13 years at a capital cost of $US2 billion. Some states have woken up (Ohio, Virginia) and have frozen or stopped renewable energy mandates. Wind power provide electricity intermittently regardless of demand, cannot provide electricity when it is needed (i.e. peak demand times) and a spinning reserve of coal- and gas-fired plants need to keep operating and releasing CO_2 to the atmosphere for when the wind does not blow.

The US Energy Information Administration (EIA) argues that onshore wind is one of the cheapest forms of electricity and is

---

631: http://eia.gov

632: http://instituteforenergyresearch.org

633: http://www.nyiso.com

634: http://newsmax.com/LarryBell/Climate-Change-Global-Warming/2015/008/03/id/665118/

cheaper that nuclear,[635] coal, hydro and solar. A report from Utah State University[636] shows that the EIA's true costs are around 48% more expensive than claimed. States have enacted Renewable Portfolio Standards (RPS) which require utilities to purchase electricity produced from renewable sources at a high cost for consumers Wind turbines are a long way from transmission lines, expansion of the grid is expensive and costs were passed on to taxpayers and consumers. Conventional electricity generation needs to be available as backup 24/7 as wind is unable to meet demand. This further drives up the cost of electricity. The Production Tax Credit alone amounts to $5 billion per year subsidy to wind producers. The real costs of wind energy don't include the environmental damage and purported harm to human health.

The Germans are rueing the day they decided to save the world and quickly convert to wind power. The German energy policy has been a disaster. Subsidies are colossal, the electricity market is now chaotic, electricity grids have problems coping with the sporadic input from wind turbines,[637] CO_2 emissions are rising quickly, German industry has decamped to other jurisdictions, more than 800,000 German homes have had their power cut off because they can't afford to pay the increased cost of electricity and the green dream of creating thousands of new jobs has turned into a cold black nightmare.

Some Germans are facing such energy poverty that they have tak-

---

635: www.eia.gov

636: http://www.usu.edu/ipe/wp-content/uploads/2015/04/Renewable-Portfolio-Standards-Colorado.pdf

637: Grid systems have tolerances of a few volts and Hertz hence any excessive variation results in a shutdown. On 13 June 2015 between 9am and 3pm, the whole South Australian grid collapsed from 750 MW to 50 MW, a drop of 94%. It's been known for a long time that feed-in solar and wind power leads to uncontrolled power surges and blackouts and attempts to increase the proportion of renewables to the grid result in more blackouts and no savings of CO_2 emissions

en to the forests to collect wood for heating and cooking. God knows how Germany will be able to provide electricity in winter if 800,000 illegal immigrants from the Middle East and Africa arrive on their doorstep. US electricity prices are about 33% of those of Germany and Obama is heading down the same path as Germany. Why?

Denmark has the most expensive electricity in Europe. When the wind does not blow, Denmark buys coal-fired electricity from Germany, nuclear electricity from France and Sweden and hydropower from Norway. We saw a screaming headline[638] telling us that *"Wind power generates 140% of Denmark's electricity demand."* However, this 140% was for only a brief moment on a windy night at 3 am, which is the time when demand is the lowest. Call up the Danish Electricity Authority site[639] and look at the map of inputs and outputs of electricity.

There is far more electricity imported to Denmark from conventional sources than is exported from wind. Denmark is not self-sufficient in electricity, as wind does not provide enough for consumers and employment-generating industry. Wind-power is inefficient costly ideology. It's not hard to see that misguided energy policies have exacerbated the economic decline of Europe. If wind power was efficient and reliable, green left newspapers like *The Guardian* would not have to be misleading and deceptive. Denmark's state-owned energy company Dong Energy has stopped building onshore wind industrial complexes because of the public outcry. At times, Denmark's wind energy cannot be used in Denmark and it is exported. The end result is that about 7% of Denmark's energy is from wind. The rest is imported.

Australia has a population of 24 million spread over an area 7,692,024 square kilometres with an electricity demand of more than 6 times that of Denmark. Denmark has 5.7 million crowded into

---

638 http://www.theguardian.com/environment/2015/jul/10/denmark-wind-windfarm-power-exceeded-electricity-demand
639 http://energinet/dk/EN/EI/Sider/Elsystermet-lige-nu.aspx

only 43,094 square kilometres and no major electricity-consuming industries such as smelters providing metals to the rest of the world. Australia is an island and we can't rely on neighbouring countries to provide us with electricity if our ideological energy schemes fail. Like Denmark, our electricity consumption is low at 3 am and our pricing is highest at peak morning and evening times. South Australia, which has a higher proportion of wind power, is higher cost than the other eastern states. What a surprise.

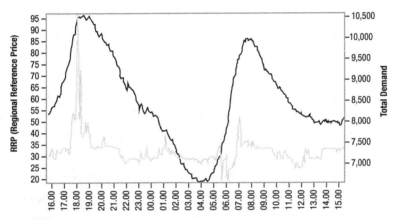

Figure 8: Peak, base load and minimum load electricity during 24 hour day in Victoria (10 August 2015); solid line is total demand, grey line is retail price.[640]

In the UK, wind turbines are subsidised. Wind turbine owners are even paid for electricity that is not used and are compensated because consumers are not using wind power. This is the surreal world that UK governments have created in an attempt to "decarbonise" the economy. The UK National Grid[641] warned that coal-fired power stations are being closed down so quickly that the spare capacity of

---

640: http://www.aemo.com.au/Electricity/Data/Price-and-Demand/Price-and-Demand-Graphs/Current-Dispatch-Interval-Price-and-Demand-Graph-VIC
641: http://www.ft.com/intl/cms/s/0/f3d1b352-fef4-11e4-84b2-00144feabdc0.html#axzz3ilYqjyfx

16% in 2012 was only 1.2% in 2015. Most will be closed by 2023 because of EU rules intended to curb "carbon" emissions.[642] What will another cold winter show? The UK is bordering on a self-induced disaster that even the thousands of costly emergency back-up diesel generators will not solve.

Renewable energy targets

EU blackouts

The EU's policy to close coal-fired electricity plants and to increase the use of renewable energy is taking away sovereignty from Britain and causing the National Grid to be at risk from blackouts and has increased the cost of electricity to consumers. This is exactly what the Obama administration wants for the US. While Europeans are learning the hard way and getting rid of subsidies, the US want American families to experience the same pain. The only pain will be increased electricity prices, a loss of reliable electricity and a loss of jobs.

Germany has scored an own goal. Germany's shift to "renewable" energy has had a greater impact on traditional coal- and gas-fired power plants than originally planned. Some 57 traditional gas and coal plants will close as a consequence of Germany's *Energiewende*. It is now almost impossible to build a new modern plant.[643] If you visit Germany, bring candles and a warm coat. The candles must be coloured green, of course.

Germany relies heavily on Russian gas. Russia is not known to be a stable supplier and uses energy, especially in winter, for political blackmail. Russia and the Ukraine have pulled a three card trick and, as a result of political horse trading during UN negotiations, most of the carbon credits created by them do not represent a cut

642: http://www.bloomberg.com/news/articles/2014-07-09/most-uk-power-plants-seen-shut-by-2023-on-climate-rules.html
643: *Deutsche Welle*, 24 August 2015

in emissions because they involved emissions derived from fires on coal mine waste dumps and flaring of gas during oil production.[644] Why am I not surprised?

UK renewables

UK Office for Budget Responsibility has some pretty chilling estimates on the effects of renewable energy targets in Britain.[645] In 2014, the environmental levies to support wind turbines, solar panels and biomass plants was £3.1 billion *per annum* and will rise to £9.4 billion *per annum* in 2020. Such policies now account for 5% of energy bills and in 2020 will account for 15% (£226 per person pa). There is a target of 30% energy from renewables by 2020. During the same time period to 2020, environmental levies will cost British households a cumulative £89 billion. Wind electricity generators receive £85 to £90/MWh for the next 20 years of which £40 is a subsidy. No one has addressed the real issue: Once the UK loses 70% of its electricity supply from fossil fuels, what will they do when the wind does not blow and the Sun does not shine in a country already starved of sunlight? The UK consumer is getting restless.[646]

The UK Climate Secretary Amber Rudd is axing subsidies to wind and solar power and protecting taxpayers from rorts (such as feed-in-tariffs for solar panels) that have spiralled out of control.[647] She said that she understands why people see tackling global warming[648]

644: http://www.sei-international.org/mediamanager/documents/Publications/Climate/SEI-WP-2015-07-JI-lessons-for-carbon-mechs.pdf

645: http://www.telegraph.co.uk/news/earth/environment//11483899/Green-levies-on-energy-bills-to-treble-by-2020-because-of-renewable-targets-official-figures-suggest.html

646: http://www.telegraph.co.uk/news/earth/earthnews/11679649/Rising-green-energy-levies-risk-public-backlash.html

647: Pilita Clark, *Financial Times*, 24 July 2015; Ben Webster, *The Times*, 23 July 2015

648: UK Government Axes Green Deal; Global Warming Policy Foundation, 24 July 2015

as *"cover for anti-growth, anti-capitalism and proto-socialism."* The Prime Minister David Cameron has moved to get rid of the "green crap" that has driven up energy bills and made businesses uncompetitive.[649] No subsidies would be offered to those who install domestic solar panels, a system that allowed wealthy families to rake in subsidies paid by poor people struggling to pay electricity bills.[650] Subsidies for renewable energy have reached £1.5 billion *per annum* as part of the government's "green deal". Minister Amber Rudd has slashed £540 million from a scheme that gave out loans and cash for energy efficient home improvements.[651]

The UK government is cutting subsidies in a move to protect millions of families from rising energy bills and the noise from rent-seekers is deafening. Energy companies, investors and even local councils were well aware that wind energy was subsidised by legislation and that a change in government or government policy was the main financial risk. Some Scottish councils invested in wind electricity projects hoping to profit from the higher electricity bills in the rest of the UK. Now they are up in arms that their snouts might be forcibly removed from the trough. These councils have not been stopped from their wind projects so they could easily pass on the additional costs to their council taxpayers. All that has been stopped is councils mining the national subsidy market.

US renewables

The first renewable energy mandates in the US were imposed in 1983. Most states did not impose mandates until about 20 years later. Over 30 states had mandated voluntary renewable energy requirements. Ohio was the first state in the US to freeze its renewable energy mandate. Utilities would have been required to provide 25% of the state's

649: "Get rid of the green crap", *Daily Mail*, 21 November 2013
650: *Daily Telegraph*, 28 August 2015
651: *The Australian*, 9 August 2015

electricity from renewable sources by 2025. Ohio halved its mandate level because of high costs. West Virginia repealed its renewable energy mandate and in New Mexico the renewable standards were frozen. Kansas is well down the path to repealing its mandate which would save ratepayers $171 million (i.e. $4,367 for each household). States with a higher mandate had higher unemployment than states without a mandate. The US Department of Energy has found that electricity prices have risen in states with renewable energy mandates twice as fast as those with no mandate. Electricity prices in states with mandates are 40% higher than those with no mandate.

The US Federal government has paid $US18.1 billion as green energy subsidies since 2006 aimed at encouraging households to install energy-efficient windows, air conditioning systems, roof top solar panels and to buy electric and hybrid cars. A study by the University of California showed that the lower income 60% of US households (by income) received about 10% of the amount of "green credits" whereas the upper 20% (i.e. with incomes above $75,000 pa) extracted 60% of the amount. For electric vehicles, the upper 20% of income earners received 90% of the green credits.[652]

Tesla Cars received $US256 million to produce electric cars and General Electric spends tens of millions in lobbying for more taxpayer's renewable energy dollars. Such cars can only be afforded by the wealthy who, because they are such good environmentally-concerned citizens, will receive $US 2,500 to $US 7,500 in tax credits (as well as free charging and express lane access).

Renewables in Australia

The Australian government is also looking at ways of stopping subsidies and the green left environmental activists, energy companies and investors are complaining with support coming from the normal

652: *The Australian*, 12 August 2015

media suspects. Subsidies in the form of higher electricity prices for consumers will eventually end and this should come as no surprise in an allegedly free market system for energy companies and investors.

Climate change has been a huge business opportunity. Green left environmental activists who seem to detest employment-producing businesses have handed a huge business opportunity on a platter to the climate change industry. With the long-established contradictory climate change policies between the two political parties, any business that takes a risk and invested in wind or solar power is taking a huge risk. If the policy changes for the correct fiscal reasons, just wait for the screams and emotional language.

Political policy

There is never a good time for bad public policy. For half a century, the green left environmental gurus have been telling us that renewable energy would soon be as cheap and as reliable as coal, gas, oil and nuclear energy. The turning point and great new discoveries are just around the corner and what is required is even more public funding and tax breaks before there is abundant, cheap, clean, reliable energy that will just materialise out of thin air.

We have waited for 50 years, the dream time is over and there are far too many examples showing that renewable energy is not environmentally friendly. Huge areas of wild lands and wildlife habitats are now industrial wastelands of wind and solar energy facilities, food-growing areas have been replaced with biofuel crops and there has been genocide of birds and bats. Apart from the high environmental costs, the economic costs have forced a rethink. In most jurisdictions, it has been a case of legislate in haste (and to win votes) and repent in leisure (and to win votes).

Sustainability

We hear a lot about environmental sustainability. I don't really know what this means as it has not been defined. However, I am all for sustainability. Fiscal sustainability. This can be easily defined as keeping out of the red. Although there may be an argument for government funds to be used as venture capital seed money for new technologies, there is a time when such monies should dry up. There were no government seed funds for 19th century inventions. Either they worked or you went broke.

The greatest period of innovation in the history of the world was driven by capitalism and government was on the sidelines and not getting in the way with over regulation and micro-management. The withdrawal of subsidies for wind and solar electricity generating schemes is a good example. There has been enough time to demonstrate that such alternative energy is sustainable and efficient and yet there is a perception that such schemes should have permanent subsidies at the expense of the consumer in case something happens well after we are dead.

If human-induced climate change is real and needs to be tackled, the best way is to use free enterprise and competition to drive down energy costs and to develop new technologies. Tried and proven inefficient unreliable old technologies such as wind and solar for electricity generation have had their day. It the 19th century, inefficient technologies that did not live up to expectations as advertised fell by the wayside.

Henry Ford's motor cars didn't become best sellers because they were subsidised, because horses were heavily taxed or because of government decree. They were cheap, efficient, reliable, faster, did the job as advertised, were better than horses and didn't fill cities full of hazardous poo and smells thereby creating an environmental problem. Wind- and solar-generated electricity are not efficient, not reliable, not environmentally beneficial, don't do the job as advertised, bad for human health and fiscally unsustainable.

Over time, Henry Ford's mass-produced cars became better whereas neither wind- nor solar-generated electricity has become more efficient. It is time to stop taxpayers' money propping up inefficient unreliable ideological electricity generation. It's called common sense and fiscal responsibility.

Making sure that each generation does not amass a monstrous debt for the next generation is also fiscal sustainability.

Rare and endangered species

The Pope shows concern for rare[653] and endangered[654] species. It appears that mining damages the environment yet renewable energy industrial sites do not. The example given by the Pope in the Encyclical of damage by mining and smelting was a method used primitive mining and smelting in Third World countries[655]. Such mining and smelting does not take place in the developed world and is driven by poverty in the Third World.

It is one of the great mysteries of the world that endemic, threatened and protected plant and animal species only live where new mining, construction and energy projects are planned but somehow don't exist where wind and solar facilities are planned. And what a surprise, when claims about the species being endemic, rare or threatened are examined by mining companies during detailed environmental impact studies, it is invariably shown that the species have a wide distribution and are not so rare after all. This just shows how little we know.

Misleading miscreants

The planned new $16 billion Carmichael coal mine in Queensland

---

653: *Laudato Si'*, Paragraph 37
654: *Laudato Si'*, Paragraphs 91, 123 and 168
655: *Laudato Si'*, Paragraph 51

will provide desperately needed 10,000 new jobs and provide electricity to 100 million people in India who have no electricity. This is no flash in the pan, jobs will be provided for decades and generations of Indians will be able to rise from poverty and benefit from cheap coal-fired electricity for at least 50 years. The Federal Court ruled for a delay in construction by the Indian company Adani because the Federal Minister for the Environment did not take into account his department's advice about the yakka skink (*Egernia rugosa*) and an ornamental snake (*Denisonia maculata*). What a load of nonsense it is to assume that these two species are restricted to that area and furthermore that they would be threatened by carefully planned development.

Heads should roll in the Department if the Minister was not fully informed. The most likely scenario is that a green left environmental activist in the Department of the Environment "forgot" to supply the Minister with the critical documents.

Is this really a matter for the Federal Court? Political and administrative matters should be dealt with in parliaments whereas courts should resolve legitimate *bona fide* legal disputes. In order to get into the Federal Court and to launch proceedings, all the green left environmental activists need to do is to show that they are environmental activists or researchers and have been active over the last two years. The rules of standing have been lowered so much to encourage green left activist litigation that if the argument is not won in front of politicians and bureaucrats, then green left environmental activists can take their argument into a more sympathetic forum: the courts. Judges are not equipped to balance the complex process of conflicting interests. This is the business for governments as is acting in the public interest whereas the green left environmental activists are quite happy to sabotage vital economic projects and undertake a political vendetta against employment-generating development.

If the Carmichael mine is stopped, this will have no effect whatsoever on emissions of CO_2 from burning coal. What it will do would be to raise the price of electricity for Indians who are already poor, raise investment risk in Australia, create more unemployment in Australia and push Australia towards a Greek-type economy. No thanks, lets give the Indians a chance to escape from crippling poverty.

A few salient facts might show how the tail wags the dog in Australia. The area for the proposed mine comprises brigalow and outcrop and is a long way from nowhere. The Wet Tropics World Heritage Area is 270 km north and the Great Barrier Reef is 200 km east or 320 km downstream via watercourses and a river. The environmental footprint of the proposed mine is 3 square kilometres. Adani started its environmental impact study in 2010, have already spent nearly $1 billion of the $16 billion to be spent and are required to fund a closure and rehabilitation strategy until 2074. No multinational octopus is tearing away at pristine wilderness and damaging the Great Barrier Reef in order to mine "killer coal."

The environmental impact statement documents were voluminous and prepared by independent scientific, environmental and engineering groups.[656] By any measure, there was a best practice comprehensive study. Public comments on these documents in early 2013 attracted 14,464 submissions, most of which were from environmental activists.[657] Of the online submissions, 36% were from NSW and 24% from Victoria. It is highly unlikely that these interstate green left agitators have ever been to the proposed mine site and their ill-informed objections could only be ideological. Only 17% of submissions were from Queensland.

656: Volume 1 68pp, Volume 2 79pp, Volume 3 85pp plus hundreds of pages of appendices plus supplementary documents written after public consultation comprising three volumes and 62 appendices.

657: 68 submissions from landholders, agencies, local government and private organizations and more than 14,000 submissions from green left environmental agitators.

After rewriting what was a very technical work, the final documents were hundreds of pages in length. The Queensland government co-ordinators' office then came in with its two bob's worth and its report was a 585-page evaluation of the final and amended environmental impact statement wherein there were notices and inputs from the Federal Department of the Environment. This is an extraordinarily expensive process costing hundreds of millions of dollars to provide jobs in Queensland and to provide poor Indians with cheap energy. It was not that Adani's proposal threatened the environment or tried to circumvent the Australian environmental regulation system. They did everything that was required of them.

The green left environmental activists tried to stop a new mine opening and ran a misleading and deceptive campaign. Statements like this from green left environmental activists should be treated under the Corporations Law. Companies must not be engaged in misleading and deceptive conduct. Why shouldn't environmental activists and unionists be constrained by the same common sense laws?

The Federal Court overruled the plans of a government that received 5,882,818 votes to win an election and supported groups that carp about the failures of the political system. The political system has failed, in the eyes of Greenpeace, because the majority of Australians don't support Greenpeace's extremism. This is not "progressive" politics as such actions condemn Indians to eternal poverty and the green left environmental activists use every coal-fired communication tool in modern Australia to agitate against poorer nations who try to achieve our standard of living.

It is clear that the courts and unelected green left environmental activist groups think that they should be making policy decisions rather than the elected officials. This disdain for democracy is anarchy. It looks very much as if the taxpayer-funded Federal Court is behaving as an activist group over-riding State and Federal interests, due process and economic policy. If a judge in the Federal Court makes a wrong decision, their job is safe. If an elected official makes a wrong decision, they could lose their job.

The environmental lawfare action was brought by the Mackay Conservation Group[658] who create no new jobs and who have no concerns whatsoever about a yakka skink. They are 600 km away from the proposed mine. They would wet themselves and run screaming back to their fossil-fuel driven cars if confronted with a snake. All they want is to stop a new coal mine on ideological grounds and claim that these reptiles will be threatened by a mine. Do they realise that any coal from the Carmichael mine emits CO_2 in India, not Australia.

A check of the Australian Reptile Online Database[659] shows that the known distribution of both the yakka skink and the ornamental snake are larger than the area of NSW and Tasmania respectively and both species were first described in the 19th century. They are not newly recognised species, not rare species and not endemic to a small area in and around the proposed coal mine.

Figure 9: Geographic distribution of yakka skink[660]

658: www.mackayconservationgroup.org.au
659: http://arod.com.au/arod/

Figure 10: Geographic distribution of ornamental snake[660]

Green left environmental activists need to be reminded that if major employment generating projects are blocked, then there will be no future revenue streams for hospitals, schools and other public services. Maybe if the Mackay Base Hospital, Pioneer High School, Mackay State High School and the James Cook University School of Medicine and Dentistry at Mackay were all closed because of a lack of future mine revenues, the message might sink in to the Mackay Conservation Group. Does the Federal Court need to be reminded that productive industry and their employees pay tax, this tax is collected by governments. Distributed tax pays for the Federal Court to keep these inner urban blinkered judges in their ivory towers away from reality.

Governments paying to stop employment

The Queensland and Australian governments encouraged Adani to push ahead with the mine. To make matters even more scandalous, the Queensland Labor Government gave $50,000 of taxpayer's money to the Mackay Conservation Group[660] who then use the money to try to close down industry and stop 10,000 new jobs in the Mackay and Central Queensland area. Furthermore, the Mackay Conservation Groups was financially supported by GetUp! and the NSW government's Environmental Defender's Office.[661]

The Environmental Defender's Office of NSW received $750,000 in the 2014-2015 financial year from the Public Purpose Fund. In 2013-2014 it received $221,885 in recurrent funding and $1.2 million from the Public Purpose Fund and in 2012-2013, the Environmental Defender's Office received $1.4 million from the NSW public purse.[662] This NSW public money has been used to successfully challenge the building of a mine in Queensland. NSW government funds used to try to stop a mine in another state that would generate revenue for the Queensland government. You couldn't make this stuff up.

It gets worse. Environmental activists were given $5 for every job they killed in Queensland by the Queensland government and yet the Queensland government 2015-2016 budget web site for Mackay is headed "Jobs now, jobs for the future of Mackay." The Labor Queensland Mines Minister Anthony Lynham insisted that the Federal Court delay would not stop the proposed project. What he did not state is that delays cost money, delays tempt companies to abandon projects because of increased sovereign risk and this delay was partially funded by his own government.

Where is the leadership from government? It appears that government says one thing and government officials do another thing. If government officials make a small mistake, they get a cuddle and take stress leave on full pay. If they make a big mistake, they get promoted to another department. In the private sector a small mistake earns a black mark and a big mistake results in dismissal.

660: http://www.budget.qld.gov.au/regional-budget-statements/mackay.php
661: *The Australian*, 17 August 2015
662: *The Australian*, 21 August 2015

Wotif founder Graeme Wood and former Greenpeace employee John Hepburn were involved in setting up a company, The Sunrise Project, to assist an indigenous group to fight Adani. This has split the indigenous Wangan and Jagalingou people in the area with some paid to stop jobs and others wanting jobs for generations. A payment of $325,000 was made "*to initiate a community development program and explore their alternatives to mining on their country*" and access to a scholarship program connected to the University of Queensland to the value of $600,000 over five years was also canvassed.[663] Not wanting to be outdone, the Green Party co-deputy leader Larissa Waters stated on ABC Radio 612 Brisbane on 25 August 2015: "*They should not start digging up the Great Barrier Reef.*" No one from the taxpayer-funded ABC corrected her.

Plutonic critters

There are already very high environmental standards in Australia and, in my experience, once a mining, construction or energy company does a comprehensive study of the environment for a planned industrial site, then rare species are found. If the study is extended to outside the proposed site, the same rare species are also found showing that they are not so rare. It's not that the species is rare, it is that a comprehensive study of the environment is rare.

For example, the rare Pilbara leaf-nosed bat (*Rhinonicteris aurantia*) was found in the Sulphur Springs area of the Pilbara in Western Australia. Old mines provided the perfect habitat. Environmental impact studies for other proposed mines have shown that the bat is widespread. Drill holes at Sulphur Springs contained stygofauna[664] communities, which at the time they were thought to be extraordinarily

---

663: *The Australian*, 21 August 2015
664: Blind unpigmented critters that live in fresh water and wet soil, derive food from organic matter in percolating ground water and are probably climatic refugees from previous higher rainfall times.

rare but since the initial finds in exploration drill holes, some 350 species have been found in more than 200 locations in the Pilbara with the total number of species estimated to be 550. Most species discovered were new to science and endemic to the Pilbara.[665] What was extraordinarily rare was in fact very common, only because exploration drilling was able to sample beneath the surface (which is not common practice by zoologists).

Stygofauna are most commonly reported from caves, they were previously known from the Cape Range, Kimberley, Nullarbor Plain and between Eneabba and August in Western Australia and were not known from the Pilbara in 1993 when a review of Western Australian stygofauna was published.[666] For those that struggle with Scrabble, stygofauna can help.[667]

If green left environmental activists actually left the couch, stopped dreaming of ways to spend other people's money and went into the never-never to undertake a comprehensive environmental study on an area, then they might learn that in wealthy countries there are very high environmental standards, that wealthy countries have the funds to address real or perceived environmental issues and many environmental problems are perceived and not actually based on measurement.

Many perceived environmental problems are actually a result of incomplete databases of species distribution and it is environmental

---

665: Eberhard, S. M. *et al.* 2005: Stygofauna in the Pilbara region, north-west Western Australia: a review. *Jour. Roy. Soc. West. Aust.* 88, 167-176

666: Knott, B. 1993: Stygofauna from Cape Range peninsula, Western Australia: Tethyan relicts. *WA Museum Records and Supplements* 45: 109-127

667: Smith, Anthony 1953: *Blind white fish of Persia.* Penguin. Smith describes the tasty fish stygofauna from the underground irrigation canals of Iran (called qanats). One fish species was named after him (*Nemacheilius smithi*). During Scrabble, if you are ever stuck with a q without a u, think of a qanat and its stygofauna. End of trivia, back to the Papal Encyclical.

impact studies by mining companies that greatly add to the data-bases.

Olympic hysteria

Who can forget the endangered green and golden bell frog (*Litoria aurea*) found in 1993 during site development for the 2000 Sydney summer Olympic Games. For a time, it looked like one frog would stop the Games. No gold medals, just a golden frog.

Ponds were built, frog fences constructed, frog friendly vegetation was planted, eleven road underpasses were built to prevent frogs from hopping onto the internal roads in the Olympic site, $6.5 million was spent restoring an old brick clay pit into a frog sanctuary[668] and the frogs all lived happily ever after at the taxpayers' expense. If St Peter is a green and golden bell frog, then I'm sure that Sydney-siders will get a front row seat in Heaven.

I wonder if the Chinese did the same for the 2008 summer Olympic Games in Beijing?

Activist fraud

Australia

The environmental activist Jonathan Moylan is an anti-coal activist and led a group called Front Line Action in Coal. He published a fraudulent media release to 295 journalists and 98 media organisations that announced that the ANZ bank was divesting $1.26 billion from the Whitehaven Maules Creek coal project in NSW. In a matter of hours, $314 million was wiped off the value of Whitehaven Coal on the stock market. When journalists rang, Moylan claimed he was the genuine ANZ employee Toby Kent. He was convicted in the

---

668: http://www.sopa.nsw.gov.au/___data/assets/pdf_file/0006/347847/Pro-tecting-and-restoring-green-and-golden-bell-frog-habitat.pdf

NSW Supreme Court, received a one year eight months suspended prison sentence on condition of two years good behaviour and a $1,000 bond. Moylan claimed he was contrite and remorseful yet his action involved a high degree of planning and premeditation. He was, in my opinion, contrite and remorseful only because he was caught.

Many investors, particularly retirees, lost a lot of money through no fault of their own. If anyone in the corporate world manipulated markets as Moylan did, they would be in gaol for a long stretch because the maximum penalty is 10 years, a fine of $765,000 or both.[669] The courts are very soft on green left activists who deliberately break the law, trespass, vandalise or engage in premeditated fraud. The Australian Securities and Investment Commission (ASIC) stated at the time that they were "satisfied with the sentence".

By contrast, Australia's biggest individual taxpayer received a $130,000 fine for submitting taxation documents late. After submission of the documents and payment of tax, ASIC decided to prosecute in order to root out "bad culture." There was no intent to deceive, manipulate markets or not pay tax. It appears that market manipulation and fraud are not part of a "bad culture" whereas paying hundreds of millions in tax and submitting financial accounts late is "bad culture".

Coal is essential for the prosperity of Australia and provides jobs and cheap energy. Furthermore, the high levels of health, education and welfare in Australia are financed by trade, principally in the resources sector. Public opinion is equivocal and has difficulty in opposing the ideological campaign to stop fossil fuels and replace with renewables. The collective of the Greens and Greenpeace; various green left "think tanks"; the predominantly green left private- and public-funded media; climate change activists in schools, universities, government research bodies, unions and political parties; wealthy individuals and the Christian churches have used constant noise, social

---

669: Section 1041E of the Corporations Act.

media and money to push for renewable energy to replace fossil fuels.

Fossil fuels currently comprise 86.5% of the global energy mix and the International Energy Association[670] predicts that they will meet 76% of the world's energy demands in 2050. In 2014, China was completing a new coal-fired power station each week, China added three times more coal-based electricity than wind and solar combined and clearly it is not getting out of coal. China is getting out of dirty coals, those with high ash and high sulphur, and getting into cleaner coal and nuclear energy.

Maybe Jonathon Moylan who thinks mining is so bad should actually do something positive for the environment. A colleague of mine who has spent his life in the mining industry started a practical environmental group in Port Melbourne. They use their spare time to collect plastic rubbish washed up on the beach. Later in Melbourne, a banker started a similar group at Albert Park and environmentalists followed suit at St Kilda and Elwood.

The Pope writes of "filth"[671] and he is quite correct. Plastic garbage is strewn over the land (especially in developing countries) and locked in great oceanic gyres. It is a huge environmental problem and each individual person can actually do something about it. This is one of the great environmental problems that the world faces. I wait for the time when Moylan organises such activity and spends his generously large amount of spare time doing something that actually makes a difference to the world environment. Anyone want to make a wager?

India

Prime Minister Modi of India knows that unless his country has cheap energy, there is no way for hundreds of millions of his people

---

670: http://worldenergyoutlook.org
671: *Laudato Si'*, Paragraphs 21 and 161

to escape from poverty. India has very large resources of coal and this could be used to pull Indians out of poverty using cheap coal-fired electricity. However, Western NGOs want India to be environmentally pure and generate electricity by wind and solar schemes. Of course, there is no consideration of the cost, inefficiency and unreliability. Greenpeace has been playing its normal games by acting against the economic interests of people and taking instructions from their foreign non-resident bosses on even trivial day-to-day administrative matters and senior appointments.

India has had enough of a large foreign anti-development corporation operating fraudulently with no responsibility and acting against its interests of India. Greenpeace had been very active in opposing genetically modified agriculture. If ever a country needed greater agricultural yields it is India. Genetically modified plants and animals are simply an improved form of the age-old practice of selective breeding. In fact, unless we go into virgin forests foraging for foods, almost everything we eat is genetically modified. India knows that the green revolution using genetically modified foods, fertilisers, insecticides and herbicides has fed the world and reduced poverty on an unparalleled scale.

The central government of India has frozen Greenpeace's accounts. The Tamil Nadu Inspector of Registrations office has issued a show cause notice on 16 June 2015 after inspectors visited the Greenpeace office on 3 June. This is the first step to winding up the Greenpeace business in India. The Inspector of Registration found major anomalies and irregularities in the office and account books.[672] Why am I not surprised?

Furthermore, Greenpeace had shifted its head office from Bengaluru to Chennai without legal permission to work in Chennai and

---

672: "Greenpeace's days in India are almost over" *Business Insider, India*, 23 July 2015

the Chennai office was not at the registered address. It appears that Greenpeace can deliberately ignore the sovereign law of India in order to protect Indians from environmentally damaging themselves. Apparently the environment is more important than the law. In India, Greenpeace have worn out their welcome.

This is a constant thread of environmentalism. They take it upon themselves to save us from ourselves. Well…no thanks. I can quite adequately look after myself, if it is all the same to you. Oh, and while you're at it, I don't need to be told how to think.

Russia

Greenpeace have form in parts of the world that have muscle. This is not very smart. On 18 September 2013, Greenpeace activists tried to scale Russia's Priraziomnaya drilling platform in the Arctic Ocean. This act of piracy endangered others in the sovereign waters of the Russian Economic Zone again demonstrates that Greenpeace are an anarchist organisation with no concern for sovereignty. The Russian authorities had earlier banned Greenpeace from entering the Russian Economic Zone.

Greenpeace ignored the ban, obviously because their high morality ideology is superior to Russian sovereignty. The Greenpeace activists and crew of *Arctic Sunrise* were lucky not to be incarcerated for life or treated energetically in true Russian style. The Russians were uncharacteristically patient and gentle.

For major new industrial developments in Western countries, there is a due process procedure that has been established by democratically elected governments. Greenpeace and any other group have the opportunity to argue their case against such developments. However, this relies on the law, logic and validated information. If the process is not to the liking of Greenpeace, there may be protracted legal challenges. However, this again relies on the law, logic and validated information.

If a decision is not to their liking, then anarchy become the order of the day with demonstrations, trespass, property damage and thuggery. Greenpeace has morphed from an organisation that hung banners from tall buildings to an international business organisation dependent upon donations from the public in order to conduct anarchy. They have discovered that climate change activism is a massive global business.

The ecologist and co-founder of Greenpeace Dr Patrick Moore has left the organisation because they no longer are an environmental group and are a far left anarchist collective. And what does Greenpeace do? They are now trying to erase him from history. Just remind me again, where has this happened before?

Ask a Greenpeace follower to give an example of a mine, smelter, factory or large-scale agricultural enterprise that is operating in an environmentally acceptable standard to Greenpeace. You will draw a blank from those who enjoy all the benefits of the modern world and the products from these industries.

The Pope needs to be careful of the company he keeps.

Gilding the lily

We hear from the media that, in the scientific community, there is little controversy and 97% of climate scientists conclude that humans are causing global warming. Is that really true? In the scientific circles I mix in, there is an overwhelming scepticism about human-induced climate change and many of my scientific colleagues claim that the mantra of human-induced global warming is the biggest scientific fraud of all time. And future generations will pay.

Furthermore, in my plus 40-year scientific career, there has never been a hypothesis where 97% of scientists agree. Just go to any scientific conference. Conferences are a collection of argumentative sceptical scientists who don't believe anything, argue about data and argue about the conclusions derived from data. There are fads, fashions, frauds and fools in science. Scientists, as well as lawyers, bankers, unionists and all other fields, can make no claim to being honest or honourable and various cliques of scientists have their leaders, followers and enemies. Scientists are no different from anyone else and they are human with all the human frailties. Scientists differ from many in the community because they are trained to think, analyse, criticise and be independent. Unless, of course, whacking big research grants for "climate science" are waved in front of them.

And what is a climate "scientist"? This is not a field of science. It is an invention of an exclusive club to exclude all those mathematicians, physicists, chemists, biologists, astronomers, geologists and meteorologists who don't follow the ideology. This club thrives off research grant cash grabs funded by taxpayers. And what does the taxpayer get in return? Climate "scientists" trying to frighten people witless so that there can be more taxpayer funded research. If one is to study climate, then almost every field of science needs to be studied and integrated. This is just not possible.

This is why the "climate science" clique, mainly comprising computer modellers and meteorologists, excludes those with the most to offer such as solar physicists, astronomers, geologists and CO_2 chemists. Climate "science" is geared to confirm the ideology of human-induced global warming and no other idea will be investigated. What about a study of the natural climate cycles? What about a study on the possibility of global cooling?

A recent paper on the scientific consensus of human-induced climate change was a howler.[673] The paper by Cook *et al.* (2013) claimed that published scientific papers showed that there was a 97.1% consensus that man had caused at least half of the 0.7°C global warming since 1950.

673: Cook *et al.* 2013: Quantifying the consensus on anthropogenic global warming in the scientific literature. *Envir. Res. Lett.* 8: doi:10.1088/1748-9326 /8/2/024024

How was this 97.1% figure determined? By an inspection of 11,944 published papers. Inspection? Is this the way that rigorous scholarship is undertaken? This was not a critical reading and understanding derived from reading every one of the 11,944 papers. This was not physically possible as the study started in March 2012 and was published in mid 2013 hence only an inspection was possible. What was inspected? By whom?

The methodology section of Cook's note tells it all:

> *This letter was conceived as a 'citizen science' project by volunteers contributing to the Skeptical Science website www.skepticalscience.com. In March 2012, we searched the ISI Web of Science for papers published from 1991-2011 using topic searches for 'global warming' or 'global climate change'.*

This is translated as: This study was a biased compilation of opinions from non-scientific politically-motivated volunteer activists who "inspected" 11,944 scientific papers, who were unable to understand the scientific context of the use of "global warming" and "global climate change", who rebadged themselves as "citizen scientists" to hide their activism and scientific ignorance, who did not read the complete paper and, if they did, were unable to critically evaluate the complexities of the science published therein. The conclusions were predictable because the methodology was not dispassionate and involved decisions by those who were not independent. If this was a financial study upon which investment decisions were made, people would have gone to gaol.

As part of an independent re-evaluation, the original 11,944 papers were read and the readers came to a diametrically opposite conclusion to Cook *et al.*[674] Of the 11,944 papers, only 41 explicitly stat-

---

674: Legates, D. R *et al.* 2013: Climate consensus and 'Misinformation': A rejoinder to Agnotology, Scientific Consensus, and the Teaching and Learning of Climate Change. *Sci. Educn* 24: 299-318

ed that humans caused most of the warming since 1950 (i.e. 0.3%). Cook *et al.* had flagged that just 64 papers supported the consensus but only 23 of the 64 actually supported the consensus. Of the 11,944 climate "science" papers 99.7% did not say that CO_2 caused most global of the warming since 1950. Not one paper endorsed a man-made global warming catastrophe. Not one! So what was the necessity to publish misleading and deceptive information? Cook's career depends on the number of papers published. This is how universities operate. Ask me. I've been there.

Furthermore, Cook *et al.* used three different definitions of climate consensus interchangeably and arbitrarily excluded about 8,000 of the 11,944 papers that expressed no opinion on the climate consensus. It appears that Cook's lackeys (i.e. "citizen scientists") made scientific judgments yet they were not scientists. The Cook *et al.* paper showed the opposite of their predetermined conclusions which can be gleaned from the Skeptical Science website.[675] Readers can make up their own minds whether the 97.1% figure is misleading, deceptive, fraudulent or all three.

The Cook *et al.* paper certainly does not look very rigorous and looks like propaganda that an unknown journal accepted. This certainly is not the scholarship expected from taxpayer-funded university staff. Cook is far from a young man, is still undertaking PhD work in cognitive psychology, he is at the University of Queensland's Global Climate Institute where he is a "Climate Communication Fellow" and runs an online course on *Making sense of climate science denial*. The course was developed "To further the work of educating the public and empowering people to communicate the realities of climate change." This looks like an imitation of Al Gore training presenters to promote the scares presented in his 2006 film *An Inconvenient Truth*. We taxpayers are funding an under qualified

675: www.skepticalscience.com

person to undertake propaganda in support of a political ideology. As Cook's paper shows, evidence is in short supply. And now for a heretical thought: Maybe those who deny that human emissions of CO_2 drive climate change use evidence.

Legates stated:

> *It is astonishing that any journal could have published a paper claiming a 97% climate consensus when on the authors' own analysis the true consensus was well below 1%. It is still more astonishing that the IPCC should claim certainty about the climate consensus when so small a fraction of published papers explicitly endorse the consensus as the IPCC defines it.*

The Cook *et al.* (2013) paper is not in accord with some of Cook's other papers. In my lifetime of science, each new publication builds on previous work. In that way, there is no great Eureka moment in science but an increasing body of repeatable validated evidence that strengthens a theory. There are also publish or perish pressures and there needs to be constant stream of publications to win new research grants and to stay employed. This pressure leads to decreasing the standard of science and contradictory publications. Cook is a prime example.

After the Cook *et al.* (2013) paper, he kept the research grant wheels spinning as coauthor of another paper on scientists' views about global warming.[676] This is not rigorous research as it is just a compilation of the opinions of 1,868 passengers on the gravy train who have a vested interest. However, the Verheggen *et al.* (2014) paper hints that the consensus among climate "scientists" on the gravy train might not be as strong as thought. Cook was a coauthor of the Verheggen *et al.* (2014) paper.

In 2013, Cook published that the consensus was 97.1% and in

676: Verheggen, B. *et al.* 2014: Scientists' views about attribution of global warming. *Envir. Sci. Technol.* 48: 8963-8971

2014 the consensus was now 43%. The authors also suggest that "*as the level of expertise in climate science grew, so too did the level of agreement on anthropomorphic causation.*" Expertise was subjectively equated with the number of peer-reviewed papers that had been published in the climate "science" literature. Blimey, those in First Class on the gravy train are experts whereas those in Third Class are not. You heard it here first. Papers published in sociology journals are always an amusing read, especially when one has a stiff drink at hand, and clearly demonstrate the dumbing down of our education system.

Such shoddy work raises many questions. How could the Cook *et al.* 97.1% suddenly drop to 43% in a year? This was not explained. The survey took place in March-April 2012 yet during March 2012 Cook was compiling data for his Cook *et al.* (2013) data. Cook would have known that his March-April 2012 data used in Verheggen *et al.* (2014) was not in accord with his March 2012 data used in Cook *et al.* (2013). Cook *et al.* (2013) did not explain the contrary results that he was acquiring in March-April 2012 for another publication.

I smell a rat. Did Cook get the idea from Verheggen, conduct his own hasty survey using "citizen scientists" and gain publication priority by beating Verheggen into print? If I ever shared the same funeral pyre as this mob, it still would be hard to warm to them. In the corporate world one can go to gaol for such behaviour and I see no reason why dodgy behaviour in the academic, political or union world should be any different from the corporate world.

So why was Verheggen *et al.* (2014) different from Cook *et al.* (2013)? Neither paper had a disclaimer by Cook. Was there a sudden change in opinion by climate "scientists", was the methodology flawed, did the "citizen scientists" skew the data or do different data sets give different answers thereby rendering the whole process unreliable? The methodology of Verheggen *et al.* (2014) was very subjective. Some 6,550 people were invited to participate in the survey and only 1,868 participated. I often get asked to participate in such surveys and the invitation always ends up in the circular filing cabinet. A participation rate of 29% does not give confidence and is not representative of the potential survey population.

Respondents were picked because they had authored climate "science" papers between 1991 and 2011 that had key words such as "global warming" or "global climate change". This is a clumsy survey because Verheggen *et al.* (2014) were not required to read the papers, were not required to understand the science and were not able to understand whether the paper surveyed was good or bad science. A computer search of key words does not give confidence in either the methodology or the conclusions.

Fabius Maximus[677] analysed the Verheggen *et al.* (2014) study and showed that only 64% of passengers on the gravy train agreed that human emissions of CO_2 was the main or dominant driver of more than half the temperature rise. This was despite the fact that there has been no measured increase in the average global air temperature for more than 18 years. Of this 64% (1,222 participants), only 797 agreed that it was *"virtually certain"* or *"extremely likely"*. That's only 43% of climate "scientists" that agree with the IPCC statement:

> *It is extremely likely (95% certainty) that more than half of the observed increase in global average surface temperature from 1951 to 2010 was caused by the anthropogenic increase in greenhouse gas concentrations and other anthropogenic forcings together.*

However, despite some pretty shoddy methodology on opinion compilation, if only 43% of climate scientists agree with the IPCC's 95% certainty, from my experience in science this is what would be expected.

Long gone are the days when experts can use their authority to state that "they know." Try that in court as an expert witness and

677: http://fabiusmaximus.com/2015/07/29/new-study-undercuts-ipcc-keynote-finding-87796/

you'll be chewed up and spat out in a few minutes. There is a huge body of literature that shows that expert opinions can be wrong. There is nothing wrong with saying "*I don't know*", as I have said to generations of students. This was done by 47% of the scientists surveyed by Verheggen *et al.* (2014).

As soon as there is a claim that 97% or 95% of scientists agree, then I disbelieve such a claim. And so should you. Many times I have had a group of geologists on an outcrop in the bush and heard the arguments. All the arguments are underpinned by the same evidence that is in right front of their very eyes. The only consensus achieved is that all participants agree that they don't agree with each other.

If there is a consensus in science, then it's not science but political activism or fraud. None of the 102 consensus computer models predicted that there would be no global warming over the last 18 years. They all predicted an accelerating warming rate. None of the 102 consensus computer models can replicate what has happened over the last decades or centuries. Yet it is exactly the same computer models that are used to predict that we will all fry and die well after the catastrophist computer modeler has died.

Rather than argue using evidence, major trillion dollar policies are argued by stating that there is a 97% consensus. The policy debate starts with assuming there is a problem with climate. There isn't. Climate is complex, we don't understand it and models that try to understand it have failed. All the calculations show that if one nation reduces its CO_2 emissions by 5% or even 50%, it will make no measurable difference to the global average temperature. Furthermore, the developing world will continue to out-emit the developed world. After all the policy fights, political battles and trauma and we eventually decide to have a carbon tax, emissions trading scheme, carbon capture and storage, renewable energy target, wind turbines, solar cells and everything else on Earth to stop using coal, will global temperatures be lowered. No.

So why do we do it? National pride at international climate conferences such as Copenhagen and Paris? Political prestige, ego, popularity and status? Greater government control of every aspect of our lives? Promotion of a global government? Shifting of more power to unelected bureaucrats, courts, academics, NGOs and green left environmental activists? Creating artificial carbon markets which both banks and socialists lust over? Destruction of capitalism? It's certainly not about the environment or the economy.

This will cost future generations dearly.

Driving jobs offshore

Up until the Industrial Revolution, humans in Western countries were beasts of burden and we laboured just simply to exist. This is still the case in many Third World countries. Yet it is green left environmental activists in wealthy Western countries that want us to go back to pre-Industrial Revolution times. They pontificate in front of a microphone using all the benefits of modern society and want us going back and being beasts of burden. Let's watch while they lead by example.

Clean green energy is expensive, even in wealthy Western countries like USA where there is abundant cheap oil, gas and coal. States in the US with renewable energy mandates are used by the anti-fossil fuel lobby to force consumers to pay more for renewable energy. Someone has to pay and it is the consumer.

For example, in energy rich Kansas, each consumer pays an extra $US 4,367 *per annum* for energy because of renewable energy mandates.[678] For a time, US industry was driven offshore because of high renewable energy mandates.

Green taxes in the UK are so high that in mid July 2015, Tata

---

678: http://www.washingtontimes.com/news/2015/mar/29/h-sterling-burnett-pulling-the-plug-on-renewable-e/?page=allhppt://www.washingtontimes.com/news/2015/mar/29/h-sterling-burnett-pulling-the-plug-on-renewable-e/

Steel cut 720 jobs (July 2015)[679] followed by another 250 jobs (August 2015)[680] because electricity costs *"are more than double those of key European competitors"*. If not changed, the EU emissions trading scheme would add £30 a tonne to average UK steel costs. The Redcar steel plant at Teeside will close after 160 years of continuous operation with 1,700 direct and 4,000 indirect jobs lost. Green taxes for CO2 emissions have put thousands out of work in an already depressed area. Electricity prices are expected to rise by 47% by 2020.[681] UK manufacturers are already moving elsewhere because the decarbonisation policy is a major competitive disadvantage to the UK.

Pope Francis calls for a lower carbon-lower-impact future but surely was not calling for higher costs and more unemployment.

How nations become healthy and wealthy

The answer is simple. Use fossil fuels. History shows us that their use is the path to national wealth, health, longevity and a better environment.

Electricity availability

According to OilPrice.com, the proportion of "clean green energy" used in the Western world is very low (2%). The main energy source is oil. It is used for transport; coal, gas, hydroelectric and nuclear for electric power generation and coal for smelting and refining. Without fossil fuels for transport, no food could come from growing areas for consumption in cities, no concrete could be used for construction and no trade could be undertaken. This is the ecomarxist green utopia that some of us oppose. The papal Encyclical also has hints of ecomarxism.

679: http://www.bbc.com/news/business-33550863
680: http://www.bbc.com/news/uk-wales-south-east-wales-34065990
681: Global Warming Policy Foundation, 30 September 2015

The International Energy Agency states "availability of electricity is one of the most clear and undistorted indications of a country's energy poverty status." There is an almost one-to-one correlation between GDP of the top 20 countries and electricity generation. Cheap electricity is an antidote to poverty and the associated poor health, environmental damage, poor education, child mortality and a short life.

Table 3: Global energy consumption in 2015[682]

Energy source	Production (billion BTUs)	2015 global production (%)
Crude oil	46,163,119	32.2
Coal	43,971,202	30.6
Natural gas	33,953,799	23.7
Hydroelectric	9,133,759	6.4
Nuclear	7,377,624	5.1
Wind	1,755,206	1.2
Biofuels	697,809	0.5
Solar	349,369	0.2
Geothermal	95,705	0.1
Total	143,497,592	

Over the last decade, global energy consumption has increased by 28% (i.e. equivalent to 56 million barrels of oil) and today some 86.5% of global energy comes from fossil fuels. Although fossil fuels have been demonised by green left environmental activists, it is clear that they will underpin the global energy mix for a very long time,

682: http://www.eia.gov

especially as coal is abundant, low cost, widespread geographically and is not constrained by an OPEC-like entity.

Renewables can't even keep pace with the growth in electricity demand, let alone displace fossil fuels unless, of course, we voluntarily all decide to become miserably poor. If the Pope and his green activists want to stop the use of fossil fuels, then economies will collapse and the world would collapse into starvation, poverty and conflict because there is no viable low cost, low capital expenditure, energy efficient alternative. Are the Pope and his green left environmental activist advisors immunised against reality?

The Second Industrial Revolution

We Westerners are beneficiaries of the First Industrial Revolution. There is a far larger industrial revolution taking place in China, SE Asia and India. This second Industrial Revolution is quicker than the first and affects far more people than the first Industrial Revolution. Nevertheless, there are still 1.2 billion people with no access to electricity. According to the World Bank, another 2.8 billion people use solid fuels such as twigs, dung and wood for home cooking and heating. These fuels emit toxic fumes that kill women and children. Yet many in the West want to stop the current Industrial Revolution. The Pope laments this situation but does not offer the blindingly obvious solution: coal-fired electricity. A few examples tell the story.

Since 1985, Thailand has increased its coal-fired electricity generation and increased its CO_2 emissions by 600%, Vietnam has increased its electricity generation by 2,500% and Indonesia has increased its coal consumption by 6,000%.[683] In 2013, Indonesia exported at least 330 million tones of coal, over the last 5 years China and India increased their Indonesian coal imports by 101 and 70 million tonnes

---

683: Bryce, Robert 2015: More energy please. *IPA Review* February 2015

respectively. Indonesia exports a third of the steaming coal traded annually and has now overtaken Australia as the world's largest exporter of coal on both tonnage and energy content.[684] These three countries have more than 400 million people, the average *per capita* GDP is $6,000, and far from wanting to use wind and solar energy and remain poor, they want more and more energy from cheap coal.

India is the world's third biggest emitter of CO_2. Just visit India. Even in the most remote parts of the sub-continent there is a pall of unhealthy polluting smoke from the use of biological fuels rather than coal-fired electricity. The Indian environment minister Prakash Javadekar is determined that India will not cut its CO_2 emissions ahead of the climate summit in Paris in late 2015. The minister stated that his government's priority was to alleviate poverty and improve the nation's economy which would involve an increase in CO_2 emissions through new coal-fired electricity and transport. When asked about cuts to CO_2 emissions, the minister stated:

> *What cuts? That's for more developed countries ... India's first task is eradication of poverty ... Twenty percent of our population doesn't have access to electricity and that's our top priority. We will grow faster and our emissions will rise.*

Javadekar stated the obvious:

> *... unless we eradicate poverty, we cannot really address climate change....to that end, we need to grow. Our net emissions must increase.*[685]

Javadekar also argued at the "World Day to Combat Desertification" that poverty is *"an environmental disaster"* and that the BASIC group of nations (Brazil, South Africa, India and China) have the

---

684: International Energy Agency, www.iea.org/newsroomandevents/graphics/2015-02-17-indonesia-coal-exports.html

685: http://zeenews.india.com/news/eco-news/india-has-right-to-grow-carbon-emissions-may-rise-prakash-javadekar-960613.html

right to grow.[686] Maybe the Pope should have taken advice based on experience and knowledge from the leaders in poor countries rather than from wealthy Western green left environmental activist atheists who show little concern about poverty. No Encyclical will stop the growth of BASIC, SE Asian, South American and African countries and it is immoral to try to stop people escaping crippling poverty.

According to the World Bank, over the last 20 years, 1.7 billion people gained access to electricity with about 800 million of these people acquiring electricity from burning coal. Over this period, 100 million Indonesians also gained access to electricity. Concurrently, per capita GDP rose by 442%, life expectancy increased by eight years, infant mortality fell by 45%, child malnutrition fell by 65% and illiteracy declined by 77%. This is not a coincidence. Indonesians have been grindingly poor, many are still poor but coal is giving Indonesians the path to escape from crippling poverty.

The Pope shows concern and compassion for the poor and is also concerned about development, economic growth and the use of fossil fuels. He offers no solution to global poverty. Maybe his advisors could have suggested that the solution is to develop and burn local coal resources to underpin a reticulated cheap electricity network. Almost every country has coal resources. If Indonesia, SE Asian countries, China and India can start to escape from poverty today, so can other countries.

The only moral position the Pope can take is to argue that poor countries should generate coal-fired electricity.

686: http://indianexpress.com/article/business/economy/don't-blame-us-for-carbon-emissions-india/

5

WHAT THE POPE SHOULD HAVE BEEN TOLD

I have great respect for the Pope's sincere wishes to end pollution and poverty. We all share the same sentiments. The solution is to use cheap coal-fired electricity and not to demonise coal and other fossil fuels. The Industrial Revolution and the growth of East Asia and India show that with cheap coal-fired electricity, people are brought out of poverty. It has happened to hundreds of millions of people over the last 20 years.

Burning coal releases CO_2. This is the gas of life. Plants are fertilised by CO_2 and use less water and there has been a greening of the Earth with the slight increase of CO_2 in the atmosphere. The food for all life on Earth has been demonised as a pollutant.

It has yet to be shown that CO_2 drives global warming and all models of future climate based on CO_2 have failed. Despite hysterical predictions based on models, planet Earth has not deteriorated due to an increase in CO_2 in the atmosphere. Nature and humans add traces of a trace gas CO_2 to the atmosphere.

The planet has not warmed for more than 18 years, models predicted a steady temperature increase over this time. A model-predicted hot spot over the equator does not exist. The models are not in accord with measured reality and must be rejected. However, the assumptions that are used to construct a model are not transparent. The science on climate change is far from settled, there is no consensus and there is no overwhelming evidence of human-induced global warming.

Furthermore, in the past when the Earth had a high atmospheric CO_2, there was no tipping point, no runaway global warming, no accelerated extinction, no increase in bacterial and viral deaths and ecosystems thrived more than today. When the past atmospheric CO_2 content was up to 1,000 times higher than now, there were ice ages, no acid oceans, no correlation between temperature and atmospheric CO_2 and no correlation between atmospheric CO_2 and sea level.

This high atmospheric CO_2 content was removed into sediments and life and eventually sequestered in sedimentary rocks. There has been no compelling case made for the reduction of CO_2 emissions by humans, models of future climate have overestimated the project-ed rate of warning and have totally ignored the possibility of global cooling. Geology and history show us that global cooling kills people and destroys ecosystems.

There is quite a bit of explaining to do if it is argued that human emissions of CO_2 drive climate change. The 0.4°C temperature rise since 1940 during a period of wars followed by intense industrialisa-tion was about the same as the previous 40 years when the emissions were much lower. This has not been explained but strongly suggests that CO_2 is a small bit player in driving climate change. Furthermore, why was the 19th century warming associated with almost no change in atmospheric CO_2, again suggesting that there is little relationship between CO_2 and global climate.

The Pope's promotion of renewable energy shows that he was not well advised. Wind, solar, wave and tidal forces do not have the energy density to keep modern society alive. Wind and solar indus-trial complexes release more carbon dioxide than they save and are inefficient, unreliable and need back up all day and all night from coal, gas, nuclear or hydro. In order to try to make renewable energy more competitive, governments have increased the costs of conven-tional electricity to the point where there is fuel poverty in Western countries and employment-generating businesses are closing down or moving.

No Third World country trying to escape from poverty can afford renewable energy and it is only Western countries that use renewable energy because they are wealthy. Wealthy countries didn't become wealthy overnight and centuries of the evolution of free trade, democracy, creativity, resource utilisation and property rights made wealth creation possible. Governments, collectives or international treaties did not create this wealth. Individuals created it. By denying poor countries access to fossil fuels, Pope Francis condemns them to permanent poverty with the associated disease, short longevity and unemployment.

In his second Encyclical, *Laudato Si*, the Pope seems to have swallowed hook, line and sinker the new environmental religion. The Encyclical is an anti-development, anti-market enthusiastic embrace of global green left environmental ideology. Much of the 40,000 word Encyclical is a denunciation of free markets. This is dressed up as religious instruction from the largest church in the world. I am sure that he is only too familiar with the Parable of the Talents (or the Parable of the Bags of Gold) in Matthew 25: 14-30.

The Pope may be trying to revive Roman Catholicism, especially amongst young people, by being an environmental populist and wants his Encyclical to be part of Catholic teaching. The Encyclical is meant to be a teaching letter with the Pope speaking as a moral and spiritual guide and not as a scientist or politician. The Encyclical is not meant to be a document of public policy yet it is.[687]

Most Encyclicals are about hope[688] whereas *Laudato Si'* is actually a depressing doomsday view of the future strengthened by the absence of evidence, science and discussions about uncertainty. The

---

687: *Laudato Si'*, Paragraph 118
688: *Laudato Si'*, Paragraph 13

Pope shows concerns for the poor yet only offers constraints that would make the poor poorer.[689] There are no scientific references in the Encyclical even though much of it is about science and it attempts to use science to make comments about the future.[690]

The Encyclical warns of "doomsday" yet it was written at a time when hundreds of thousands of Christians in the Middle East are facing actual doomsday. Christians are suffering calculated slaughter, beheadings, torture, abduction, slavery and rape by Islamists. This is the real doomsday about which the Pope writes nothing in the Encyclical.

The Pope has only listened to a small group of green left environmental activists, some who are in a warm embrace with communism. The Pope was very poorly advised. He clearly did not consult eminent people with contradictory evidence and conclusions. He did not consult conservative Christian scientists. The Encyclical provides support for the one thread that runs through the whole green left environmental activist movement. It is deceit. Whether it is in scientific methodology, scientific publications, media reporting or political spin, deceit is always present. If an ideology cannot be promoted without deceit, then it should be abandoned. The Encyclical will survive the test of time but for the wrong reasons.

Green left environmental activists have shown that they are mean treacherous, shameless liars who are vulgar, cowardly, anti-environmental and ignorant. Their embrace of the principles of communism should send shivers down the spine of anyone who has a sense of history or who lived through these times (including millions of European Catholics). Communism as an ideology is anti-democratic and anti-human and is totally opposed to Catholicism. The green left environmental activists are anti-industry, anti-capitalism, anti-ad-

689: *Laudato Si'*, Paragraph 52
690: *Laudato Si'*, Paragraph 23

vertising, anti-selling, anti-property, anti-profit, anti-patriotism, anti-monarchy, anti-empire, anti-police, anti-armed forces, anti-A bomb, anti-authority and anti-Christianity. They are against almost everything that made the world a freer, safer and more prosperous place.

Politicians want us to believe they are saving us from ruin. Religious leaders such as the Pope want to reinforce original sin and the need for repentance. Green left environmentalists see industrial progress as sin and that repentance is gained by giving up all of the benefits of the modern world. Some business leaders want the taxpayer to subsidise their expensive unproven green technologies. Climate "scientists" want their taxpayer-funded money-making machine to keep on giving. The Pope had a chance to show some leadership. He failed.

Global living standards have improved, people are wealthier, fewer people live in abject poverty and more people have access to sanitation, clean water and electricity. Although there is still a lot to achieve, the toll from diseases has decreased, people live longer, fewer people are killed from extreme weather events and there has been no increase in economic damage from extreme weather events. All in all, the world is a better place. A slight increase in CO_2 in the atmosphere had increased crop yields and has increased forest area and productivity. The net impact of a slight increase in atmospheric CO_2 has been beneficial to the biosphere.

The Third World and the developing countries desperately need to escape from poverty. The Pope's concern for the world's poor will amount to nothing unless they can have safe drinking water and affordable and reliable electricity for heating and cooking. No longer should the poor die from the smoke emitted by burning dung, leaves and twigs in huts.

Only when Third World children can do homework at night using cheap coal-fired electricity can they escape from poverty. By not having cheap coal-fired electricity, there is no electricity to pump water and treat sewage. Separate reticulated water and waste water systems have saved more lives on Earth than any other invention.

For now and the foreseeable future, only fossil fuels and nuclear energy can provide affordable and reliable power for the whole world.

The Pope should have been told the truth and should have searched for the truth himself.

CPSIA information can be obtained
at www.ICGtesting.com
Printed in the USA
BVHW041522130120
569179BV00008B/63/P